I0820144

Frontera Freeways

Highway Building and Displacement in El Paso, Texas

Miguel Juárez

Number 13 in the Al Filo, Mexican American Studies Series

University of North Texas Press
Denton, Texas

Printed in the United States of America.

10 9 8 7 6 5 4 3 2 1

Permissions:
University of North Texas Press
1155 Union Circle #311336
Denton, TX 76203-5017

The paper used in this book meets the minimum requirements of the American National Standard for Permanence of Paper for Printed Library Materials, z39.48.1984. Binding materials have been chosen for durability.

Library of Congress Cataloging-in-Publication Data

Names: Juárez, Miguel, 1958- author
Title: Frontera freeways : highway building and displacement in El Paso, Texas / Miguel Juárez.
Other titles: Al filo no. 13.
Description: Denton, Texas : University of North Texas Press, [2025] | Series: Number 13 in the Al filo, Mexican American studies series | Includes bibliographical references and index.
Identifiers: LCCN 2025037052 (print) | LCCN 2025037053 (ebook) | ISBN 9781574419795 cloth | ISBN 9781574419870 ebook
Subjects: LCSH: Express highways--Design and construction--Social aspects--Texas--El Paso | Highway planning--Social aspects--Texas--El Paso | Mexican American neighborhoods--Texas--El Paso--History | Minorities--Relocation--Texas--El Paso
Classification: LCC TE25.E96 J83 2025 (print) | LCC TE25.E96 (ebook)
LC record available at https://lccn.loc.gov/2025037052
LC ebook record available at https://lccn.loc.gov/2025037053

Frontera Freeways is Number 13 in the Al Filo, Mexican American Studies Series.

Parts of this book are based on Miguel Juárez, "From Buffalo Soldiers to Redlined Communities: African American Community Building in El Paso's Lincoln Park Neighborhood," special issue, *American Studies* 58, no. 3 (2019): 107–24.

The electronic edition of this book was made possible by the support of the Vick Family Foundation.

Typeset by vPrompt eServices.

To my wife, my siblings, our late parents, and all persons displaced by highway building.

Contents

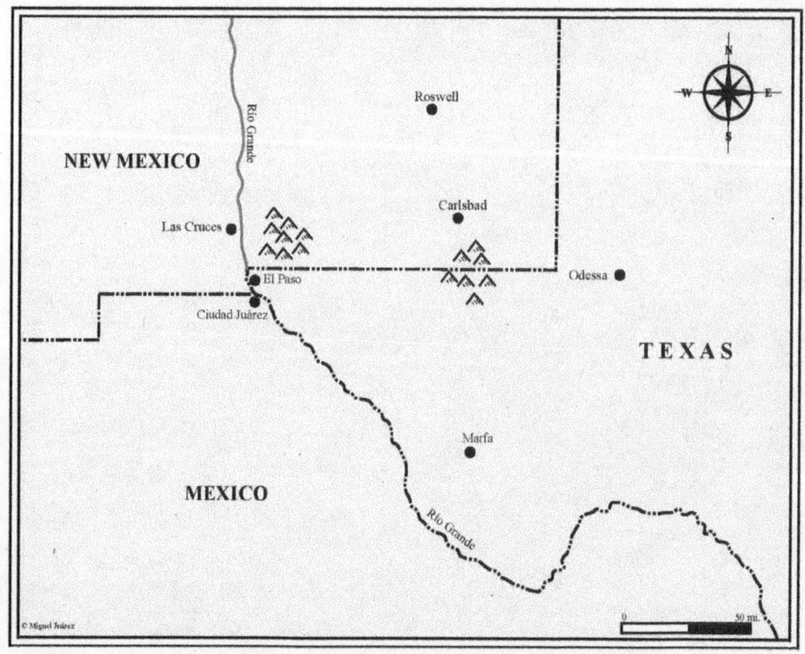

Figure 1 El Paso, Texas, and Cd. Juárez, Chihuahua, Mexico. Map by Alex Mendoza.

Preface

The effort to save an over one-hundred-year-old schoolhouse in South Central El Paso, Texas, helped to uncover the history of how the city's freeways were built, why these structures were placed in certain neighborhoods and not in others. The sources in this work reveal that the official story is based in documents, both primary and secondary sources, including issues of race, class, politics, and funding. The unofficial story that communities often hear is based on incomplete stories or secondhand information, innuendo, and/or rumor. It then becomes the work of historians to tease out the unofficial story with facts and with proof, correct it, and reveal what has been silenced, looking at the silences or facts that have never become known, so that we can then confirm the true history that has occurred.

Aside from the creation of El Paso's first freeways and the displacement of thousands of citizens in processes that were unfamiliar to them, the story here is how an El Paso, Texas, Lowrider Car Club named the Latin Pride Car Club reinvented and rebranded itself to become the Lincoln Park Conservation Committee (LPCC) to celebrate Chicano/a culture and work with the community and local leaders to save Lincoln Center in South Central El Paso from demolition.

Since 2005 Lincoln Park Day has been held every third Sunday in September; members of the LPCC would see the Lincoln Center building closed off to the community and they began wondering, What would it take to reopen it? By 2010 they began asking why it was not open and why it was closed in 2006. As the years went by and Lincoln Park Day became a larger event, the community felt it was time to stop asking those questions and start taking actions to reopen the building. They began writing letters to elected officials and initiated a grassroots campaign to bring awareness to the reopening of the center. They went it alone and met with limited success until they were joined by a key individual. If that person had not joined LPCC, they certainly would have lost their fight.

There existed the official story that Lincoln Center had been closed due to mold damage because of the rains of 2006. The community sensed that the city's story was not true. Along the way various persons and elected officials joined in the effort to reopen the center. It took a village to save Lincoln Center, despite many times the community felt it would be demolished.

This research proves that grassroots organizations working with supporters across the board can win these types of efforts. It just takes the right combination of people to come together and think things out, to strategize and not make hasty decisions. The historical analysis of the Lincoln Park community and highway building uncovered a deeply technical and controlled process that in present day is linked to federal dollars. Gone are the days when highway councils decided on the future paths of freeways. Today, it behooves highway builders and politicians to make their processes transparent to the communities they serve, or they run the risk of losing funding and/or delaying projects.

Communities across the country may not know the histories of how highway projects were developed, who decided, and how communities that were initially spared may be facing their removal or relocation today. One way to fight future projects is to research your community. Urban and history scholars can help. We have been trained to ask tough questions and support communities in finding those answers.

Most of the archival materials used for this research were readily available, including older Texas Highway Department materials in the state archives, which few had used. A challenge in engaging in this research is that older records may not be available, but records can be requested via a Freedom of Information Act (FOIA) request, but if the documents are extremely sensitive, they may be redacted. Unfortunately, state highway departments do not create finding guides for their collections or electronic files for community use.

Adding to the issue, the reality is that older electronic records are often deleted. Since more documents are born digital and are outside the reach of researchers, there need to be policies in place to make current materials from state highway departments available for research purposes. State highway departments are complex institutions with their own cultures and policies.

There remains a dearth of information and archives which are not available or do not exist, except in private collections. One of the more interesting records not available for this research included The El Paso Builders Association (EPBA) records or archives. No one could verify if EPBA records existed or were available in a private collection. An inquiry by the author into the EPBA archives produced no materials. Current officers were unaware of any materials or archives from the association's early years.

Several experiences come to mind in the creation of this work. First, I remember walking in Segundo Barrio (El Paso's Second Ward) as part of my first graduate urban history course in spring 2010 with Dr. Adam Arenson. He was the first and the last professor to teach urban history at the University of Texas at El Paso. I took two classes with him before he moved on to become a professor of history and director of the Urban Studies Program at Manhattan College. I also took a course under Dr. Yasuhide Kawashima, where I researched eminent domain in El Paso from 1880 to 1920. This research laid the foundation for this study. Another experience I had was when Dr. Max Grossman and I were looking at the exterior walls of Lincoln Center while he was doing an architectural assessment of the building, and right there and then the Lincoln Park story became my research topic.

Another experience is from October 27, 2012, when I was sitting in the first row in the audience of The Interstates, the City, and Highway Policy panel at the Sixth Biennial Urban History Association Conference at Columbia University in New York City. The session was chaired by Mark Rose, Florida Atlantic University, and panel members included Roger Biles, Illinois State University; Edward Muller, University of Pittsburg; the late Raymond Mohl, University of Alabama at Birmingham; and Michael Fein, Johnson & Wales University. Owen Gutfreund, Hunter College, City University of New York, was the moderator. I saw this research as a continuum of what this group of urban historians had researched and presented. A few years later, in April 2016, I attended the Urban History Association Luncheon at the Organization of American Historians Annual Meeting in Providence, Rhode Island. At the gathering, I sat across the table from Dr. Mark Rose, and he asked me about my work. After I shared what I was doing, he commented, "You're following in Raymond's [the late Raymond Mohl] path." All these combined experiences forged this research.

Acknowledgments

I would like to thank Dr. Howard Campbell, chair of the Sociology and Anthropology Department at the University of Texas at El Paso, who encouraged me to initiate my doctoral studies, and Dr. Paul Edison, former associate professor in the History Department at the University of Texas at El Paso, who guided my research. I also want to thank my dissertation committee, which consisted of Dr. Campbell; Dr. Bradley J. Cartwright, and Dr. Yolanda Chávez Leyva (from the History Department); Dr. Max Grossman (from the UTEP Department of Art); and Dr. Davarian L. Baldwin (from Trinity College in Hartford, CN), whom I met through the Organization of American Historians Travel Award in 2016, and when asked kindly agreed to serve on my committee.

I would also like to thank my professors who supported my work: Dr. Charles Ambler, Dr. Adam Arenson, Dr. Ernesto Chávez, Dr. Selfa A. Chew-Meléndez, Dr. Keith Erekson, the late Dr. Maceo Dailey, Dr. Will Guzman, Dr. Gary Kieffner, Dr. Cheryl Martin, Dr. Charles Martin, Diana Martínez, Dr. Ignacio Martínez, Dr. Manuel Ramírez, Dr. Julia Schiavone Camacho (Goshen College), Dr. Jeffrey P. Shepherd, and Dr. Michael Topp. A special thank you goes to Dr. Louis Woods from Middle Tennessee State University for sharing his insights on the Home Owners' Loan Corporation and the Federal Home Loan Bank Board regarding El Paso, Texas. His work regarding redlining was a crucial piece of my dissertation since it proved divestment in the communities studied in this research.

Thanks also goes to my activist-scholar friends who make up Raza Organize and the Chicano History and Preservation Project: Joseph "Clavo" Martínez; friend and attorney Ray Eli Rojas, whose article "Under I-10: Life in Lincoln and La Roca" in *Newspaper Tree* initially piqued my interest in this topic and who has been a tireless supporter of my work; and Rosemary Ibarra, who was a great early supporter of this research. Ray was also instrumental in filing two important Freedom of Information Act requests with the City of El Paso to shed light on decisions regarding the planned demolition of Lincoln Center. Special mention goes to researcher and supportive friend and community scholar Laurie Cooper.

I would also like to thank the members of the Latin Pride Car Club, who early on figured out they would get more support from city, county, and state officials if they rebranded themselves the Lincoln Park Conservation Committee (LPCC). In addition, numerous individuals such as the Honorable County Judge Alicia R. Chacón and her daughter, Corinne Chacón; Hector Gutíerrez Jr.; community members like Rosa Guerrero; El Paso City Representatives Lily Limón and Eddie Holguin; and dozens more who supported our efforts against the demolition of Lincoln Center. The community is also deeply indebted to former State Senator José Rodríguez and to former State House Representative Joe C. Pickett, who supported efforts against the demolition of Lincoln Center since 2011 and kept viable the possibility of reopening Lincoln Center. El Paso City Mayor, the Honorable Oscar Leeser, was also a big supporter of the effort to save Lincoln Center.

I would also like to thank the following narrators who allowed me to interview them for their remembrances of Lincoln School, Lincoln Park, highway building and El Paso in general: the late Manuel F. Aguilera, Joe Cardoza, Carlos Callejo, the late Tom Diamond, Joan Cunningham Estrada, Gabriel Gaytán, Hector González, Mateo and Lydia Hinojosa, Martha Hinojosa-Arriola, the late David Prieto, Dalia "Dolly" Prieto Rivero, Francisco Prieto, the late Ramon Renteria, the late Alex and his brother Robert "Bobbie" Rosas, H. L. Scales, the late Mrs. Oralee Smith, Rebecca Valdez Sterling, Cynthia and Federico Villalva, Henry Williams, the late Linda Zavala, and Jackie D. Hoyt. A special thank you to Arturo and Vallarie Enriquez of Vantage Point Photography who loaned me their copy of their video of the late Elvira Villa Lacarra Escajeda, founder of the Chamizal Civic Association, who fought the local powers for her barrio. Arturo and his parents were also displaced by the Chamizal Settlement.

Thank you to the National Association of Chicana and Chicano Studies (NACCS) Tejas FOCO, under the aegis of Jaime Armin Mejía PhD, associate professor of English at Texas States University, who saw this work as important to the Mexican American experience in Texas.

In addition, I would like to thank the other members of Senator Jose Rodríguez's Twenty-Ninth District Heritage Tourism Advisory Committee, including past chairs attorney Carmen Rodríguez and Gary Williams, who provided ongoing contextual discussions via our monthly meetings on the importance of documenting El Paso's history. Prince McKenzie also acted as an invaluable resource on El Paso's early railroads. A special thank you goes to the members of the El Paso Railroad and Transportation Museum. Thank you to researcher and curator Joseph Longo for providing me with his work researching the Clardy Fox neighborhood.

Anne Cook from the Texas Department of Transportation (TxDOT) Photography Library in Austin spent time discussing internal publications with me and provided excellent aerial photographs of El Paso. She also made time to discuss the history of the Texas Highway Department, which became TxDOT. Archivist Tony Black at the Texas State Library suggested I look at Frank Halla's dissertation and at other state highway collections; Allan Fisher, archivist at the LBJ Library, provided me with an orientation for the use of the Lyndon B. Johnson Papers and identified additional materials; and Margaret Harman and Jacqueline Thornburg provided their time for me to review LBJ's photographs of his El Paso visits in the Photography Archives. Finally, Head of the C. L. Sonnichsen Special Collections Department at UTEP Claudia Rivers and Head of Access Services / Technology Abbie Weiser, along with Eva Ross, a volunteer who processed the El Paso Planning Collection, and of course the late Juan Sandoval III—all these great people deserve my thanks for their assistance in locating important primary documents.

I would also like to give a shout out to my friends who encouraged me along my path: Cesar Caballero, Luis Chaparro, Roberto "Robb" Chávez, Mary O. Barnard, Katherine Brennand, B. J. Myers, Rick Hernández, Joe B. Rodríguez, Gaspar Enriquez, Rebecca C. Hankins, the late Dr. Roberto C. Delgadillo, Oralia Garza de Cortez, Dr. Dolores Gross, Dr. Alana de Hinojosa, artist and educator Lupe Casillas Lowenberg, the late Ramon Renteria, artist Roberto Salas, historian Maria Eugenia Trillo, and the late Stephen T. Vessels. Also, a big thank you to the countless people who donated their time, talents, and spaces to raise consciousness and facts to save

Lincoln Center: Francella Baca, Fred Borrego, Jud Burgess, Texas State Representative Cesar Blanco, Carlos Callejo, Norma Priscilla Chavez, Dr. Mariana Chew, members of Chicano Park in San Diego, Cemelli de Aztlán, the late Pete Duarte, the late Randell David Fleet, Guillermo Glenn, David González, Luis Hernández, Isela Laca, Hal Marcus, Xavier Miranda, Aaron Montes, the late Hector Montes, the late Paul Moreno, the late Enrique Moreno, Celia Muñoz, Sito Negrón, the late Jesus Ochoa, Arlina Palacios, Michael Patino, Georgina Perez, Dr. Cynthia Renteria, Ana Reza, Abel Rodríguez, El Paso County Judge Ricardo Samaniego, El Paso County Commissioner David Stout, the late poet Donna Snyder, Cuahtemoc Villegas, photographer Federico Villalba, and countless others.

I would also like to thank my mentors who encouraged me to stay on the path: the late Marta Amaya Arat, the late Alberto Bonilla, the late historian Elias Bonilla, Michael Mills, the late Ruth S. Ortego, and the late Dr. John Ellison. Lastly, I would like to thank the countless El Pasoans and media who, when needed, stepped up to the plate to help save Lincoln Center from demolition and to report on developments to keep the El Paso community informed: *El pueblo unido, jamas sera vencido* (the people united will never be defeated).

Dr. de Hinojosa, an assistant professor at Texas State University in San Marcos, has written extensively on the Chamizal Settlement and on Escajeda. In November 2024 she led a meeting in El Paso, Texas, to develop a historical marker to honor Escajeda's contributions to save her community.

A special thank you to Barbara Angus, former curator of the Cavalry Museum and the El Paso Museum of History. Mrs. Angus arrived in El Paso in 1977 and committed herself to researching, promoting, and highlighting El Paso's local and regional history.

Abbreviations

AASHO	American Association of State and Highway Officials
BOTA	Bridge of the Americas
CDBG	Community Development Block Grant
EPCC	El Paso Community College
FHA	Federal Housing Administration
HOLC	Home Owners' Loan Corporation
LPCC	Lincoln Park Conservation Committee (or Latin Pride Car Club)
LULAC	League of United Latin American Citizens
MACHO	Mexican American Committee on Honor Opportunity
MEChA	Movimiento Estudiantil Chicano de Aztlán
MPO	Metropolitan Planning Organization
MUA	Multiple Use Agreement
NAACP	National Association for the Advancement of Colored People
NACA	National Association for Chicano Arts
NACCS	National Association of Chicana and Chicano Studies
NHPA	National Historic Preservation Act of 1966
NRA	National Recovery Administration
PARD	Parks and Recreation Department (City of El Paso)
Project BRAVO	Building Resources and Vocational Opportunities
ProNaF	Programa Nacional Fronterizo (National Border Program)
TxDOT	Texas Department of Transportation
UTEP	University of Texas at El Paso

Introduction

Emergence of the Lincoln Park Community

In 1915 the first national coast-to-coast highway, the Bankhead Highway, was built through Texas, as it was in all the Southern states. It was not an accident that the nation's first coast-to-coast highway was built in the Southern states. Before Interstate 10 the Bankhead Freeway, or Highway 80, ran through Texas. In West Texas it followed the contours of the US/Mexican border, which was the Rio Grande.

In West Texas close to El Paso County, at McNary, the route dipped south and became Alameda Street. The freeway was named after John H. Bankhead of Alabama, who had been chairman of the Post Office and Post Roads Committee of the Senate. The legislation to build Highway 80 was introduced by Chairman D. W. Shackleford of the House Roads Committee on June 6, 1916, as H.R. 7617 and Senator Bankhead acted as its major sponsor in the Senate. The act was signed by President Woodrow Wilson on July 11, 1916 and it created Highway 80, which would later act as the foundation for the transcontinental freeway, the precursor to Interstate 10.[1] When it was built in the late 1960s, Interstate 10 through El Paso followed not the Bankhead Highway route but instead what was known as the "old highway," built through what were known as the "sand hills."[2]

West of El Paso, Highway 80 entered New Mexico and continued through the Southwest and into California, ending up at the Pacific Ocean. We must ask ourselves what labor force built the Bankhead Highway. It was not Chinese labor, because via its enactment in 1882, the Chinese Exclusion Act limited immigration to the United States. It was not Mexican labor, because Mexicans were in the middle of fighting in the Mexican Revolution or fleeing Mexico. In the Southern states, convict labor was used to build road projects like the Bankhead Highway.[3]

There are countless stories on how highways were built in other cities and how communities (mostly communities of color) were displaced, but this is one of the first studies to detail that history in El Paso, Texas, a city on US-Mexican border. In this work I center the Lincoln Park neighborhood, which up to the late 1960s was at the edge of the city limits. Lincoln Park originally began as the Village of Concordia, and I use this community to anchor the history of highway building in the region. It is at Lincoln Park where all of El Paso's early freeways converge; thus, the community is the focal point of this study.

This research draws on several areas: urban history, borderlands history, El Paso history, and transportation history. It also explores sociological aspects of race and the built environment in a borderlands context and incorporates urban environmental history informed by borderlands literature with a focus on African Americans and Mexican Americans who lived next to each other in the same neighborhoods, on historical memories of these communities, and on art and social protest as a form of community engagement and mobilization. Finally, it is also a study of power and how that power was used to reshape the city, which continues to this day.

Part of the challenges associated with highway building is that its processes are often shrouded in technical terms. The terminology of building roads and highways is often filled by technical jargon and terms that the general population may not know or understand, such as the concepts of "right-of-way" and eminent domain. According to the Merriam Webster Online Dictionary "right-of-way" has different meanings. The dictionary's definitions include: "(1) a legal right of passage over another person's ground; (2a) the area over which a right-of-way exists; (2b) the strip of land over which is built a public road; (2c) the land occupied by a railroad especially for its main line; (3a) a precedence in passing accorded to one vehicle over another by custom, decision, or statue; (3b) the right of traffic to take precedence; (3c) the right to take precedence over others."[4] "Right-of-way" sounds better than "the acquisition of private property to build roads and highways." It is important to note that state transportation agencies have entire departments dedicated to the acquisition of private property to build roads and highways (right-of-way). Railroad companies had similar departments when they began building our country's railroad tracks. It is the intent of this book to not only shed light on those processes but also show how sometimes how people fought back against highway building to preserve their communities and cultural heritage.

This study is timely today because due to the creation of the city's midcentury freeways, the fabric of the small town of El Paso was forever altered using processes few citizens know about, and now over fifty years later (from 2021 to today), a similar process is repeating itself and will affect other historic communities that were initially spared. Unfortunately, in the process of building highways or other transportation projects, by the time the public is invited for comment, highway decisions are done deals. Highway infrastructure decisions and actions are largely developed by paid consultants who have little or no connection to the cities being impacted. As Robert D. Bullard, Glenn S. Johnson, and Angel O. Torres have pointed out, "Transportation decision-making—whether at the federal, regional, state, or local level—often mirrors the power arrangements of the dominant society and its institutions."[5]

Highway building in El Paso, as in other cities around the country, disrupted many communities and lives, and the processes were unknown to the displaced. Individuals whose properties were taken usually received a letter and a highway representative visit. At the same time, the processes of highway building also created new and affordable suburbs in El Paso's East Side along the path of Interstate 10 and the other freeways that were constructed prior to 1970.

Emergence of the Lincoln Park Community

The Lincoln Park community originated from a village named Concordia, three miles east of present-day El Paso, in 1840. Concordia later became Lincoln Park, which was the center of major infrastructure changes that affected the region. El Paso as a borderland city became associated with expanding nation-states that encroached on people's lives and properties and displaced them in the name of modernity. The history of El Paso's freeways, like other parts of the country, demonstrates how racism shaped the process of urban highway planning in the region over a period of less than twenty years, and how city planning and transportation at local, state, and federal levels—including federal programs such as the Federal Highway Act of 1956, urban renewal, and Lyndon B. Johnson's War On Poverty—orchestrated these changes.

Interest on researching highway building grew out of the activism of the Lincoln Park Conservation Committee (LPCC), a grassroots organization that worked to counter the demolition of Lincoln Center (formerly Lincoln Park School) beginning in 2009. Lincoln Park School was built in 1916, not long after the neighborhood had been incorporated into the city of El Paso. With the coming of the highways, the school closed in 1969. Lincoln Park School reopened in 1977 as the Lincoln Park Cultural Arts Center, which first provided offices for cultural and social agencies and then for the Parks and Recreation Department (PARD). In 1985 Artist Paul H. Ramirez and I co-coordinated the *Juntos 1985* First Invitational Hispanic Art exhibit at the gallery inside Lincoln Center, which featured ten artists from El Paso and ten artists from Cd. Juárez. The exhibit led to the creation of the Juntos Art Association.

Oral history interviews were conducted with both African American and Latino/a former residents from Lincoln Park and surrounding communities, engineers and planners involved in highway building, former tenants of the Lincoln Cultural Arts Center, and artists who aided in the reconstruction of the vibrant life of the community. This work places El Paso in step with national stories on highway building but does so within a transborder, multiracial context. It contributes to borderlands and public history scholarship as a case study to describe both the transformation of the villages that found themselves north of the Rio Grande after the signing of the Treaty of Guadalupe Hidalgo and, later, how parts of the United States found themselves in Mexico after the signing of the Chamizal (Convention) Treaty in 1964 by President Presidents Johnson and Lopez Mateos.[6]

In the 1800s Chinese men worked in restaurants and laundries and lived in El Paso's First Ward. After the arrival of the railroads in the 1880, thousands of Mexicans lived alongside a sizeable Black population who worked for the railroads. The book *Porous Borders: Multiracial Migrations and the Law in the US-Mexico Borderlands* documents that "for black, Chinese, and Mexican men and women, the El Paso–Juárez border held a similar attraction of time, offering multiple peripheries that exposed the limitations of segregation laws, exclusion policies, and capitalistic desires for a captive workforce."[7] All these groups were segregated to various neighborhoods.

When African Americans came to the El Paso–Cd. Juárez region, they were employed in numerous service occupations. Historian Will Guzman in his book *Civil Rights in the Texas Borderlands: Dr. Lawrence A. Nixon and Black Activism*, writes, "Not only did African Americans find work as barbers, laundresses, housekeepers, janitors, schoolteachers, and mail carriers, but, as the town grew into an important transportation center, many Black families arrived as railroad employees."[8] In cities like El Paso, Texas, African Americans and Mexican Americans experienced discrimination and segregation via racial and class covenants and redlining. At the same time, residents of those communities lacked the effective political power to mobilize and fight against the destruction of their communities.[9] It was also an era when cities engaged in slum busting in minority neighborhoods, often toward furthering the goals of urban renewal tied to governmental funding.[10]

The Lincoln Park community was one of the first mixed African American and Mexican American communities in Texas and in the Southwest.[11] It began as a place named Concordia, or Stephenson's Concordia Ranch, in the 1840s and is located just north of the Rio Grande. With the signing of the Treaty of Guadalupe Hidalgo in 1848, following the US-Mexican War, it became part of the expanding United States. After the discovery of gold in California and continuing through the Civil War, Concordia became a respite for travelers and hosted a mercantile store for wagon trains heading West.

In 1868 part of Concordia was leased to the US Army and became Camp Concordia, or the third Fort Bliss. Garrisons of Buffalo Soldiers were stationed there. After the arrival of the railroads in El Paso in 1881, the community provided a home for both African American and Mexican workers and their families.[12]

The Mexican Revolution, which began in 1910, brought an influx of people. A nineteenth-century outpost had evolved into a twentieth-century cluster of subdivisions (East El Paso, and the French, Lincoln Park, and Government Hill additions) whose historical and cultural significance have been overlooked. This tight-knit multiracial community, like similar communities in El Paso, was deeply affected by the Great Depression. The ethnic neighborhoods faced deportation. As people abandoned their lives in the United States and repatriated to Mexico, their properties were undervalued. Nevertheless, residents found ways to survive and take care of their families.

Concordia was historically a crossroads for travelers heading north on the Camino Real de Tierra Adentro and west after the Civil War and during the Gold Rush. Later it became the Lincoln Park community, which was the site of the confluence of Interstate 10, US Highway 54, and Highway 110 leading to the Bridge of the Americas (BOTA), as well as a freeway interchange referred to locally as the Spaghetti Bowl. The development of the city's initial highway system disrupted and damaged the vibrant cultural, social, and economic structures of both Black and Mexican American neighborhoods and businesses. In his introduction to his book *From the Pass to the Pueblos, El Camino Real de Tierra Adentro National Historic Trail*, Historian George D. Torok tells us that "El Camino Reall de Tierra Adentro, the Royal Road of the Interior, was a 1,600-mile braid of trails that led from Mexico City, in the center of New Spain, to the provincial capital of New Mexico, on the edge of the empire's northern frontier."[13] By 1968 Interstate 10 and Highway 54 ran through the heart of Lincoln Park, changing its fabric and that of adjacent neighborhoods, including swaths of the Old East Side, which constituted Black El Paso. The Chamizal Freeway and the Chamizal (Convention) Settlement (1963–1964) displaced thousands of people in El Paso.[14] In addition, the creation of a large reservoir north of Interstate 10 tore down several additional blocks of residences.

Highway building in El Paso erased multiethnic communities on a large scale. This study also creates unique opportunities to examine transborder urban issues in technology, transportation, economics, and politics. El Paso's unique transborder location created Concordia and El Paso's fragmented communities along the US-Mexico border, including Segundo Barrio, Rio Linda, the Chamizal, Cordova Island, Weber, San Lorenzo, and other neighborhoods along the Rio Grande. This research follows the arrival of the railroads in El Paso in 1881, which changed Concordia's agrarian economy with the growth of the cattle industry in proximity to the US-Mexico border in a period of militarization when cattle, cotton, and copper were king.

Historical events forged the identity of the Lincoln Park community and Lincoln Park School, which itself was a site of Americanization for the neighborhood's Mexican students. Events such as the Great Depression created anti-Mexican sentiments throughout El Paso and nationally, and it spurred a decline of homeownership as thousands of Mexicans were deported or left the United States on their own. Redlining stigmatized these neighborhoods, and they would be slow to recover from it, if at all.

This work also presents the behind-the-scenes events by officials and decision-makers at federal, state, and local levels who shaped El Paso's urban landscape via the creation of the city's highways. The Federal Highway Act of 1956 became the financial conduit to build the nation's freeways, but decisions were in the hands of white chamber of commerce members, developers, politicians, and business boosters.

Highway building also coincided with suburbanization during the Cold War, as communities decentralized due to nuclear attack scares. City planning departments worked hand in hand with state and federal agencies as highway building became a mandate. Citizens did not have a voice in deciding where freeways would be built, but when possible they challenged the appraisals of their properties. This work also represents a rare view on the reasons why the elites sought to build El Paso's highways, which mirrored how decisions occurred in other cities. It also presents the costs of building El Paso's freeways and land acquisition, appraisals, planning, zoning, and ownership patterns and explores specific legislation that occurred at federal and state levels to channel millions of dollars to El Paso to build its early highways.

It presents the creation and evolution of the LPCC and their activities working with stakeholders and politicians in their efforts to save a one-hundred-year-old building from demolition. It also details the development of the Lincoln Park community as a space defined by the people who have lived in the area and continue to claim Lincoln Center and Lincoln Park as a historic heritage site.

PART 1

Before the Highways

Chapter 1

From Concordia to Lincoln Park, Three Miles East of El Paso (1800–1925)

This chapter contextualizes the Lincoln Park community by discussing its roots as the Village of Concordia, a ranch / trading post established by Hugh Stephenson and Juana Maria Ascarate de Stephenson beginning in 1840. Following the arrival of the railroads in 1881, Concordia's agrarian economy continued to grow as the railroads fueled growth in cattle less than a mile from the US-Mexico border, and it was affected by militarization and stockyard economies in a period when cattle, cotton, and copper were king.

Concordia was the product of Hugh Stephenson marrying Juana Ascarate. Together, they created Concordia as a self-sufficient community in a rural/urban setting. Later, in the early twentieth century, the Concordia heirs sold parcels of land to create housing additions and El Paso's early suburbs. Prior to 1870 Concordia's pioneers defined its space; after 1870 national politics changed its use; and after the 1880s, transportation transformed its economy. New modes of transportation altered the community, which went from an agrarian to an industrial one.

Humans have been active in the area for many millennia. Archaeologists have found "footprints—unquestionable evidence of a human presence—at white Sands National Park in New Mexico, dating to between 23,000–21,000 years ago."[1] In what later became the El Paso–Cd. Juárez region, humans followed the Rio Grande as they traveled from north to south and into the Southwest. This was the same path that railroads took when they began scouting their paths in West Texas and during westward expansion. Later in the late 1800s, railroads followed the natural course of the Rio Grande. The Bankhead Highway, built in 1915 in El Paso, also followed the path alongside the Rio Grande. Later, when the route for Interstate 10 was being considered, it also followed the Rio Grande and the railroad tracks.

Without roads Native American made their way into the present-day El Paso, Texas, region thousands of years ago. Dr. Marc Thompson and Historian Fred M. Morales, writing in a 1998 archeological report, commented that native persons made their way into El Paso and the southern New Mexico area during the Paleoindian Period (ca. 10,000–6,000 BC); the Archaic (ca. 6,000 BC–AD 100); the Jornada Mogollon (ca. AD 100–1400); the Mesilla Phase (ca. AD 100–1100); the Doña Ana Phase (ca. AD 1100–1200); and the El Paso Phase (ca. AD 1200–1400).[2] Early people used Native American trails to look for food and game and later to build shelter. The Keystone Heritage Park located on Doniphan Drive in west El Paso and the Keystone Dam archeological site (also known as Site 33 and Texas Archaeological Site 41EP494) contain some of the oldest human-made structures identified in the US Southwest, dating back more than 4,500 years."[3]

Concordia in the Mexican Period, 1821–1848

In 1840 Juana Ascarate de Stephenson and Hugh Stephenson acquired nine hundred acres of the Ascarate Land Grant; initially known as the settlement of Stephenson, the land was subsequently named Concordia Ranch and was situated on the north bank of the Rio Grande. This settlement was one of the first of several communities to develop north of the Rio Grande in West Texas; it would become the future neighborhoods of Lincoln Park and the French Addition.[4]

Hugh Stephenson, who was born in Kentucky in 1798, arrived in the region in 1824 and settled in El Paso del Norte (later Ciudad Juárez). As one of the first Anglo-American settlers in the region, Stephenson was a trapper and trader who later became a miner and merchant. In 1828 Stephenson moved to Corralitos, Mexico, to manage a silver mine belonging to Juan and Eugenia Ascarate, wealthy merchants from El Paso del Norte and owners of the Ascarate Land Grant, a large expanse of land (measuring thirteen thousand acres) awarded to the Ascarates by the Spanish Crown during the 1750s in appreciation for their military service.[5] That the same year, Stephenson married Juana Maria Ascarate, daughter of Juan and Eugenia. After acquiring the nine hundred acres from Juan and Jacinto Ascarate, Stephenson and his wife established Concordia.[6] Its boundaries correspond to present-day Montana Street to the north, the Rio Grande to the south, Stevens Avenue to the west, and Marr Street to the east (see fig. 2).

When the United States won the US-Mexican War (1846–1848), the Treaty of Guadalupe Hidalgo established the Rio Grande as the new international boundary. According to the late Jeffrey Marcos Garcilazo, "The annexation by the United States of Mexico's northern borderlands shaped the socio-economic relations between two people and cultures."[7] The treaty caused the disruption of space all along the new US-Mexico border. Historian Oscar J. Martínez writes, "Under the terms of the Treaty of Guadalupe Hidalgo, the United States compelled Mexico to accept the 1845 US annexation of Texas and forced it to cede California, Arizona, New Mexico, Colorado, Nevada, Utah, and parts of Wyoming, Kansas, and Oklahoma."[8] The Rio Grande became half of the new international boundary, with the rest consisting of an irregular line from El Paso del Norte (present-day Cd. Juárez) to the Pacific Ocean.

Concordia and all other settlements north of the Rio Grande became part of the United States, and daily activities for people were impacted with the creation of a new border. Furthermore, in 1864 the Rio Grande changed course, leaving part of El Paso del Norte in the form of a peninsula intruding in the United States. This peninsula, near Concordia, remained a point of contention between both countries until it was settled with the 1964 Chamizal Settlement.

In 1852 Josiah Frazier Crosby acquired part of the Ponce de Leon Grant and established Coon's Ranch (three miles west of Concordia), and Anson

Mills subsequently platted it as part of the El Paso township in 1859. Meanwhile, in 1854 Juana Maria Ascarate de Stephenson petitioned the Catholic Church (in Chihuahua) to establish San José de Concordia el Alto, the first private church north of the Rio Grande. At the same time, the Stephensons designated the boundaries of the Concordia Cemetery.

According to the late educator Cleofas Calleros, the church was served by "the parish priest of Our Lady of Guadalupe mission in [Ciudad] Juárez." That a priest from the Juárez mission traveled to Concordia to offer services indicates the church served a congregation of considerable size. Calleros writes, "In 1881, it was abandoned because of the objections of the Juarez priests to a Catholic burial for suicide."[9]

A later rendering of the 1852 map by the Joint Boundary Commission shows the American settlements of Hart's Mill, Coon's Ranch, Magoffinsville, Stephenson's Ranch, and La Isla (see fig. 3). According to historian W. H. Timmons, these settlements "and the proprietors played roles of great importance in the development of the area."[10] The road connecting all these settlements followed the natural terrain, as would the paths of the railroad and the highway.

According to Frank Louis Halla Jr., by the early 1850s Stephenson's Concordia Ranch was comprised of "several flat clay buildings and was distinguished by a chapel in the mode of a Spanish mission, with a sole Catholic edifice on the north bank, the ranch was focus of a substantial community."[11] Halla adds that "forty families and almost two hundred souls tenanted Concordia, all of whom either farmed small plots or labored for Stephenson and the other three merchants, relatives and friends of the owner, who made it their residence."[12] Concordia acted as a rest stop for travelers and wagon trains heading west.

In contrast, according to the *El Paso Morning Times* in 1916, "When General Anson Mills first came to El Paso there were only three stores in the place, and about forty or fifty Americans living here."[13] Concordia was comparable in size to Coon's Ranch, which was later renamed Franklin and then El Paso in 1852. John Russell Bartlett, who was contracted by the United States and Mexican Boundary Commission and who wrote a personal narrative about his experiences, described the dwellings in 1854. In his narrative he described El Paso as where many houses were built of adobe. To state that the city was a separate "country," he described the floors as built with mud, concrete, or brick and not having wooden floors.[14] Bartlett went on to state that glass windows were nonexistent except in finer homes, fires were seldom lit aside from cooking, there were a few respectable Spanish families and no middle class, and after describing the principal village of Magoffinsville, which was also the headquarters for the Boundary Commission during their stay in El Paso, he says, "A mile further east is a large rancho belonging to Mr. Stevenson [*sic*], around which is a cluster of smaller dwellings."[15]

The Stephenson children, particularly Benancia de Stephenson and her husband Captain Albert H. French, and later her second husband County Judge J. B. Leahy, would have a role in shaping the fate of Concordia, as well as what is now South Central El Paso. On February 6, 1856, a pet deer Juana Ascarate had raised as a fawn gored and killed her. She was first buried next to San José de Concordia el Alto Church but was subsequently moved to the family plot after her grave was vandalized.

Concordia During the Civil War Era

Stephenson had been sympathetic to the South and purchased Confederate bonds during the Civil War. Due to the passage of the 1862 Homestead Act, which forbade land ownership by anyone who had supported the Confederate Army, Stephenson lost his land holdings.[16]

According to historian Martin Donell Kohout, there are two accounts of how Concordia Ranch was repurchased for the Stephenson heirs: "One source says that these properties were thereupon purchased by Stephenson's son-in-law, Albert H. French, while another says that Stephenson's old friend William Wallace Mills repurchased them for Stephenson with money from the Corralitos mines."[17] Historian Nancy Gonzalez presents a counternarrative asserting that Stephenson's sons-in-law Captain Albert H. French and James A. Zabriskie were able to reacquire the land using the power they were entrusted with by the federal government in Washington, DC.[18]

In 1868 the Magoffinsville military post was flooded. Captain Albert H. French leased part of Concordia to the federal government to create Camp

Concordia (the third Fort Bliss). The camp was established over the old Rio Grande riverbed known as Rio Viejo, where water often gathered, causing the troops to contract malaria. Dwyer indicates the barracks were located where the Mitchell Brewing Company now stands.[19] Robert Neal Blake writes, "The Fort was known as Fuerte Azul because the outside woodwork was painted blue," and it was located at the site of McElroy Packing.[20] In 1868 an elementary school that would later become Lincoln Park School was created in Camp Concordia to educate the children of officers.

According to the State Historical marker for Site Concordia and Fort Bliss: "When the U.S. Army returned to this area after the Civil War, conditions proved undesirable at the prewar post, Fort Bliss at Magoffinsville. In 1868 the garrison moved to this location, then part of the Concordia Ranch. The new post was named Camp Concordia. Two barracks and other adobe structures were built. In 1869 the camp was renamed Fort Bliss. Despite poor living conditions it remained active until 1876. Troops were then withdrawn, only to return within a year. This is the third of six sites of U.S. Army posts in the El Paso area."[21] In his application for a Texas State Historical Commission marker in 1981, Thomas Moore Carson delineates the boundaries of Camp Concordia as "present-day Stevens, Frutas and Bowie Streets and the southern edge of the Concordia Cemetery."[22]

Allan W. Sandstrum describes the 1868 Concordia Post as "Paso del Norde [*sic*] (present-day Cd. Juarez), Mexico, is about 3-1/2 miles from the Post . . . the Post consists of three large buildings, (of adobe brick) two used as Men's Quarters, and one as a storehouse of Quartermaster Commissary Stores and for Officer's Quarters."[23] Other buildings necessary for the post were built by 1870.[24] Sandstrum describes the fort as having a vegetable garden irrigated by an *acequia* (an irrigation canal); grass had to be brought from fifteen miles away; mesquite bush provided the fuel for heating and cooking; timber was scarce and required a one-hundred-mile journey to obtain; water was hauled in barrels from the Rio Grande.[25] Camp Concordia was abandoned in 1877 when the military opened a new post at Hart's Mill. A caption accompanying a 1948 article in the *El Paso Herald-Post* featured a drawing of the Concordia Post and stated, "Concordia post was located in east El Paso south and east of the present Concordia Cemetery; remnants of some of the old buildings were still to be seen; the old Concordia post chapel ruins were in the Zieglar stockyards."[26]

The army base of Camp Concordia created in 1868 became home to many African Americans. Welborn J. Williams in his thesis "The Buffalo Soldiers' Brush with 'Jim Crow' in El Paso," states "Immediately after the Civil War, two companies of the 125th United States Colored Troops had been stationed at Fort Bliss." "And in the intervening years between the Civil War and 1900, the 25th and 24th Infantry Regiments and the Ninth Cavalry Regiment had served in El Paso."[27] By 1877 African American infantry troops were well-established in the region and Blacks thrived in El Paso compared to in other Southern cities.[28]

Blacks flourished in El Paso because upon their arrival there was a community they could immediately join. Organizations included the Union Aid Society, designed to improve the welfare of the community; the Knight of Pythias Lodge; and a Negro masonic chapter. Churches were also part of the welcoming fabric in the Black community. The Second Baptist Church was established in 1893.[29] According to noted historian Quintard Taylor, "No group in black western history has been more revered or more reviled than the Buffalo Soldiers, the approximately 25,000 men who served in the US Army's Ninth and Tenth Calvaries and Twenty-fourth and Twenty-Fifth Infantries between 1866 and 1917."[30] Williams states that members of the 25th were often sent to the frontier to ensure peace and stability and that their missions "were confined to stringing telegraph wire, escorting stagecoaches, building and repairing roads, and on occasion, fighting Indians."[31] In El Paso, Williams writes, Black troops were the norm due to a more tolerant racial atmosphere there than in other parts of the state.[32]

The end of the Civil War also brought former Confederate soldiers (including some who were African Americans) to El Paso. They traveled as wagon train escorts on their way to California and were required to take an oath of allegiance to the United States. A manuscript written by Willis Newton, who was 80 years old in 1921, details the story of traveling and escorting a wagon train to California from Texas in 1865. According to Newton, "When they reached Franklin (El Paso) Texas, all the grown men (which includes the older teenagers) were required to sign an oath of allegiance at the Provost

Marshall's office."[33] By the 1880s African Americans migrated to El Paso via the railroads. They worked as train workers in boiler rooms, as porters, and as assistants to mechanics. African Americans lived in Segundo Barrio, and a sizeable number moved into the Lincoln Park community.

An essay written by the late historian Maceo Crenshaw Dailey Jr., titled "Border Black: The El Paso Story," situates the African American presence in West Texas within the urban and rural borderlands and suggests how it may have differed from that of other Southern states.[34] In El Paso African Americans did not encounter the racism that was prevalènt in other parts of the country, or in other parts of Texas for that matter.[35] In West Texas African Americans found a climate of acceptance. Dailey states that "in migrating to the border region and specifically El Paso, individuals of African descent and African Americans sought their flight from oppression to opportunity, and in their new region of settlement endeavored to build communities and add their voices to the chorus for reform and positive change."[36] In the border positive change was possible, and West Texas represented a place where African Americans could be themselves, practice their religion, and not fear oppression as in other parts of the country.

Scholar Gerald Horne writes that in 1832 German visitor Carl Christian Beecher stated that Mexico's Blacks "are free in the Republic of Mexico, which is to say, they enjoy, intimately, the same rights as do the rest of the inhabitants of the state, which is not the case in the United States of North America, where, because of laws, or for prejudices, the blacks are humiliated and pressed down to the category of the lowest level of men."[37] Likewise, Dailey writes, "I am also using the term 'border black, to connote the choice presented to African Americans of the region to accept or reject their heritage and culture (avoiding alienation and deracination) and to behave based on a positive self and cultural image of their African roots and African American possibilities."[38]

The Civil War brought to the region both Buffalo Soldiers and Confederate African American soldiers who had to pledge their allegiance to the United States after the end of the Civil War so they could escort wagon trains to California. Some African Americans in El Paso and Ciudad Juárez married Mexican women. Concordia continued to grow as the railroads fueled growth in cattle, copper, and demographics and created an agrarian economy. The history in this chapter asks us to ponder why Concordia landmarks like Lincoln School and the remnants of Camp Concordia were preserved and survived into the twentieth century, but other buildings like Concordia's Casa Grande (or Big House, as the Stephenson trading post was known as) and San Concordia el Alto Church and the adobe homes surrounding the Big House were lost.

The Changing Urban Landscape, 1880–1920

With the arrival of the railroads in 1881, El Paso, having become the county seat in 1883, was transformed from a small town to an emerging city with a population of more than ten thousand by 1890.[39] Rebecca Browning Coffin arrived in El Paso from Paris in 1883 as a 19-year-old, two-week-old bride with her husband Cameron O. Coffin. Mrs. Coffin was the daughter of Confederate Colonel John Browning. She described El Paso as a frontier town with one two-story hotel (the Grand Hotel) where all the homes were built of adobe.[40] "It was such a woebegone looking little adobe town, flanked by an endless desert, that I wept when I saw El Paso," Mrs. Coffin stated in an article in the *El Paso Times*.[41] But in just four decades the face of the city would change dramatically as efforts were undertaken to modernize it. Eminent domain was one of the tools used to shape the young city and as a component for progress in the country's industrial era.

The acquisition by governmental agencies of private property from 1880 to 1920 sheds light on issues regarding property rights in a period when the built environment and the social landscape began to change dramatically. The years from 1880 to 1920 witnessed changes in the way laws and cases involving land and land titles were handled. The legal structure went from a system conducted in Spanish by Mexican lawyers to an Anglo-American system controlled by white lawyers.[42] This helped accelerate the transfer of power from Mexicans to Anglo-Americans. The use of eminent domain on citizens was as much a racial process as it was an economic one.

According to Joseph M. Cormack, "The principle underlying eminent domain has become crystallized in the form of expression set forth in the Fifth Amendment to the United States Constitution and that private property should not be taken for public use, without just compensation."[43]

Cormack adds that the use of eminent domain for land taken for individual private property or public use has been one of the most universally recognized principles of justice. "The concept of eminent domain is centuries old, and is found in the Roman law, the Code of Napoleon, and in the legal systems of the American colonies," he adds.[44] All states have adopted provisions of the Fifth Amendment in their state constitutions.

Daniel B. Bendow, in his article "Public Use as a Limitation on the Power of Eminent Domain in Texas," states that much of the property rights in existence today originate from the previous Texas Constitutions. Bendow states that the Article I, Section 17, of the 1845, 1861, 1866, and 1869 Texas Constitution stated that "no person's property shall be taken, damaged or destroyed for or applied to public use without adequate compensation."[45] After 1869 the state created the Texas Constitution of 1876, which included the Bill of Rights and section "o" that stated, "Property rights shall be maintained inviolate, except that by the "right of domain," the State has the power to appropriate any property for public use by giving reasonable compensation to the owner."[46]

As the city grew, Mexican workers immigrated to the United States to help build El Paso. The Mexican Revolution of 1910 brought an exodus of people into El Paso's Second Ward. As the city grew in the early twentieth century, Mexican and Mexican American men, women, and children worked in construction, in agriculture, on the railroads, and as domestic and laundry workers. Some opened restaurants and grocery stores that served the people of the Second Ward. In El Paso most of the Mexicans were employed as laborers and had little to no political power even though they constituted most of the population. Lacking the political or economic power to defend themselves, Mexicans were easy victims. Utilizing the concept of "right-of-way," the city was able to make a case to acquire property under eminent domain and pay next to nothing for the purchase of their properties.

Arrival of the Railroads

The United States saw the growth of regional and national railroads from 1850 to 1890. Texas officials were anxious to build railroads. As part of their land grant program, officials would give the companies a certain amount of land as incentive to build. In 1880 Concordia played a significant role in providing land for rail lines. Between 1870 and 1890, state policy forced landowners to deed right-of-way to the railroads. Concordia lay in the designated rail corridor and land was deeded for railroad purposes (see fig. 2).

According to Surveyor Terry Cowan, "A total of 35,777,038 acres of land, mostly in West Texas, was granted to 43 railroad companies."[47] Cowan states that in 1857, a corridor was created for railroad construction and "the Memphis, El Paso and Pacific Railway won the concession from the state."[48] He states that "The 'T&P' underwent several bankruptcies and reorganizations, but finally completed the railroad across Texas, meeting an east-bound tract from California at Sierra Blanca on December 16, 1881."[49]

In 1880 El Paso was a sleepy town, yet with the arrival of the railroad in 1881, it blossomed into a major transportation hub, as historian Ricardo Romo writes: "Once the train came to El Paso, the city became the hub of an inland empire composed of north and west Texas, southern New Mexico and eastern Arizona." Romo adds that "compared to other Texas cities, the economy of El Paso was quite diversified, with refineries, construction, and commerce employing the greatest portion of the work force."[50]

In comparison to Concordia, which was a resting spot for travelers embarking on their cross-country travels, El Paso became a central transportation and economic hub. Romo describes the transformation of El Paso as an industrial center, served by a regional transportation system that facilitated the delivery of products and that was supplemented by a large pool of labor.[51] In May 1881 the Southern Pacific's Chinese track-laying crews arrived in El Paso, Texas, making it a major rail crossroads, with direct access to routes connecting the Gulf Coast and Atlantic tidewater.

The railroads also helped develop El Paso's local economy and turn it into a transportation and economic hub from the east to the west and, later, from the south to the north. In addition to the Southern Pacific, other rail lines arrived in El Paso, They included the Atchison, Topeka, and Santa Fe in June 1881; the Texas and Pacific Railway, also in 1881; and the Galveston, Harrisburg, and San Antonio rail line in 1883.[52] The four railroad lines were joined by the Mexican Central Railroad from Mexico City in 1884.

The years after the arrival of the railroads saw El Paso incorporate several of its early additions: the Campbell Addition, the Magoffin Addition, the Satterthwaite Addition, the Cotton Addition, and the Mundy Addition. Hovious writes, "Directly east of the Cotton Addition between the mountain, the river, and approximately present-day Piedras Avenue was the 320-acre tract of land known as [the] Bassett Addition." She states that "Joseph Magoffin in 1880 sold that tract to O. T. Bassett, a Midwesterner most recently from Fort Worth."[53]

The railroads also helped grow the cattle and copper industry in both El Paso County and Cd. Juárez. In her dissertation on El Paso and Cd. Juárez as a cross-border metropolitan economy, economics scholar Belinda Román states that prior to 1848, communities along the US-Mexico border engaged in "small-scale agricultural production and ranching and occasional interregional trade, which allowed the communities to remain self-sufficient."[54] Historian Mario García writes that "*The Lone Star* [newspaper] commented in 1885 that in a nine-month period more than 60,000 head of cattle, horses and sheep had been transported through El Paso."[55]

Román states the cattle industry operated as a cross-border market in the United States and Mexico with relations as early as the 1840s which continued to 1920.[56] Historian Rachel St. John, in her essay "Divided Ranges: Trans-border Ranches and the Creation of National Space along the Western Mexico-US Border," describes the history of trans-national ranching: "... cattle were intentionally introduced in the US-Mexican borderlands, as part of an effort to transform the desert grasslands into a landscape of profits."[57]

Sanborn Fire Insurance maps, which were used to protect assets, from 1908, illustrate that homes and businesses in the Concordia community were still predominantly constructed of adobe and that residents heated their homes with petroleum.[58] The community was at the end of the city limits and included semi-rural features such as the Union Stockyards, a *molino* (a corn mill), and a dairy.

A part of the trans-border nature of commerce, the ranching and cattle industry along the Western Mexico-US Border defied definitions of border sovereignty. St. John states, "the suppression of raiding and the arrival of the railroads provided prerequisites for the borderlands to become part of the cattle boom that swept across western North America in the late nineteenth century, integrating rangelands across the continent into a market-oriented ranching economy."[59] In El Paso, ranching and the cattle industry boomed, and numerous stockyards were built next to railroad lines, such as the El Paso Union Stockyard in Concordia. According to the El Paso Mission Trail Association, the Union stockyards began on August 26, 1902.[60] Cross-border urbanization dramatically changed the population with the arrival of the railroads.[61] As the economy grew, so did the population, as well as the industries surrounding cattle.

In 1915, according to Orndorff, "thirty-two acres were converted into stockyards by the El Paso Union Stockyard Company."[62] In 1927, "more than one hundred thousand cattle were received in the El Paso Union Stockyards, the McElroy Union Stockyards, and various other smaller concerns."[63] In addition to cattle, Lane's Dairy was established in 1930 by Mr. and Mrs. Fitch Lane with 400 cows on Reynolds Street.[64]

Concordia Becomes Part of El Paso

A Texas key map of 1915 demonstrates the neighborhoods formerly belonging to Concordia that became part of El Paso. The nine hundred acres of land Captain French purchased at a fire marshal sale became known as the French Addition.[65] Located north of the Lincoln Park Subdivision and south of the Government Hill Addition, it was registered in June 1906. The survey of the subdivision was registered with El Paso in March 1907 and was a portion of E. R. Talley Survey No. 7.

The Lincoln Park community was an emerging neighborhood in the early twentieth century in El Paso's East Side at the end of the city limits. Like African Americans, Mexican Americans could not move into other neighborhoods because of their incomes, so they lived in the outskirts of the city or at the end of the city limits where housing was more affordable, alongside Mexicans. Mexicans migrated north to El Paso and other areas after the Mexican Revolution broke out in 1910. El Paso's various neighborhoods like El Pujido were created in proximity to the railroad. (Pujido has various references: a town called El Chinaco in Villagran, Guanajuato, Mexico, or a mountain in the Sierra Madre Occidental in Mexico, or the sound someone makes when exerting physical strain.) Like Lincoln Park, El Pujido was an ethnically mixed neighborhood, as was south El Paso.

The Lincoln Park Subdivision was a smaller part of Concordia and was located south of the French Addition. It was built south of Concordia Cemetery near Camp Concordia and next to Road 20, which later became Interstate 10. The subdivision was bordered by Concordia Cemetery to the north, Copia Street to the west, Raynor Street to the east, and the Southern Pacific railroad tracks to the south. In 1909, the Lincoln Park Realty and Import Company registered the Lincoln Park Subdivision with the city clerk's office. The Lincoln Park Addition was in Concordia School District No. 2. In 1915 Lincoln Park Realty listed 975 deeds sold.[66] Concordia Cemetery was previously larger than its present 52 acres. Concordia was half a mile northeast of the Bassett Addition, on the outskirts of El Paso, and it stopped being a village and became part of El Paso's East Side. Since Concordia was the end of the city limits, it attracted immigrants, African Americans, and other international groups. Concordia moved from being a village and became a vibrant area.

Lincoln Park: An Emerging Neighborhood

At the turn of the century, farmland dominated the landscape in the neighborhoods at the end of the city limits. With the availability of water with the passage of the Reclamation Act on June 17, 1902, farmland blossomed in the region. The Reclamation Act brought much needed water to El Paso for the development of the agrarian economy. With the development of farmland, Mexican workers moved into the community.

By the early twentieth century, Anglo-Americans outnumbered people of Mexican descent in the City of El Paso by two to one. In 1910 thirteen thousand people who were born in Mexico resided in El Paso.[67] The Mexican Revolution would change the city.[68] Historian Mario García states, "As increased fighting broke out in the north-central parts of Mexico between 1914 and 1916, thousands of Mexicans entered El Paso." He adds, "In one June week in 1916 immigration officials admitted 4,850 Mexicans into the city."[69] This number was added to the thousands already living in El Paso in 1916.[70]

Both the shortage of workers during World War I and the Great Migration facilitated the movement of African Americans out of the rural South; some relocated to El Paso. Lincoln Park was at the edge of the city limits, and numerous African Americans moved there and to Segundo Barrio.[71] El Paso Census records show slight increases of the African American population from 1900 to 1970—2.9 percent in 1900, 3.7 percent in 1910, 1.7 percent in 1920, 1.9 percent in 1930, 2.6 percent in 1940, 3.2 percent in 1950, 2.15 in 1960, and 3.12 in 1970.[72] As of July 2019, the Black or African American population in El Paso was 3.8 percent.[73]

The 1920 El Paso City Directory listed five Black churches in the city, two of which were in the Lincoln Park community: Mount Zion Baptist Church, formerly at 3129 Durazno (Rev. William Green, pastor), and Phillip's Chapel at 301 Tornillo Street (Rev. T. C. Cook, pastor).[74] Dr. Lawrence Nixon created an El Paso chapter of the National Association for the Advancement of Colored People (NAACP) in 1914. It was the first branch in Texas.[75] Historian Will Guzman's analysis of member addresses reveals that many NAACP members lived in the Lincoln Park community, on streets that were later removed due to the creation of Interstate 10.[76] Many NAACP members worked as laborers and railroad sleeping car porters. No doubt African Americans experienced racism in El Paso. Historian Ann Gabbert writes that "by 1922, El Paso's Frontier Klan No. 100 had a hardcore membership of at least fifteen hundred, including former mayor Tom Lea, and they were able to sweep the school board elections." After its short-lived two-year existence, the Klan disintegrated into opposing factions but still managed to elect members to the El Paso School Board.[77]

Lincoln Park School (The Old Lincoln School)

Concordia School first opened as a one-room school in Camp Concordia in the Officer's Quarters in 1868. Details of the activities of Lincoln Park School were found in a small journal written by Lillian E. Scott, who had been a teacher at Lincoln School and passed away in 1975. The journal included notes about her years associated with the school. The journal was found by Mary Bowling, who had been a student at Lincoln School in 1929.

According to Lillian E. Scott, who later taught at Lincoln Park School for seven years, "It was furnished with long rough wooden tables and benches, and its pupils were children of military personnel." Before long, "wisdom prompted the removal of the school to a one-room adobe building constructed nearby."[78] In 1880 the school expanded to a four-room adobe building, and at the turn of the century it reopened as a one-room brick building on Grama Street near the Franklin Canal. Its first school board consisted of "Captain A. French [co-owner with his wife, Benancia Stephenson French, of the land the school stood on], Mr. Colbert Coldwell [great-grandfather of the late Judge Colbert Coldwell, past president of the El Paso County Historical Society], and Mr. P. E. Dunne."[79]

After the creation of the Lincoln Park neighborhood from 1909 to 1915, Concordia School District No. 2 purchased several lots on Martínez Street to build a new school. The red-brick building had a basement and thirteen rooms, but because it was opened before completion, classes had to be held at the nearby dairy.[80] Dr. Max Grossman, former vice chair of the El Paso County Historical Commission, stated that Lincoln School was built between 1916 and 1917 by the local architectural firm of Buetell & McGhee.[81] The thirteen-room building was constructed of reinforced concrete and has survived.[82] On July 11, 1922, the trustees of Concordia Common School District No. 2 (S. C. McVey, Charles R. Foster, and Charles J. Mapel) sold the school to the city of El Paso for $7,469.10.[83] Lincoln Park School was built at the same time Grandview (later renamed Rusk) School was built; they are exact replicas of each other. Rusk School, located at 3601 N. Copia, was named after the Republic of Texas hero and US Senator Thomas Jefferson Rusk. It was built in 1916. On February 5, 2016, the school celebrated its one hundred years with a day of festivities.

In 1923 Lincoln Park School was incorporated into the El Paso Independent School District. It was primarily a school for Mexican children, but Black students also attended. Many of the Black students lived in the Lincoln Park neighborhood. During that period, El Paso's schools were categorized by their neighborhoods, which tended to reflect different income levels. Paul W. Horn, in his 1922 study on El Paso schools, stated, "It may be observed in passing that while these races are not segregated by any order of the school board, they tend largely to segregate themselves on account of the districts in which the children reside."[84] His report states, "Is it right or wrong for the administration of the school system of El Paso to keep in its mind this idea of two different cities, 'North of the tracks' and 'South of the tracks?'" Since Mexican Americans and "native born Americans lived in different neighborhoods," according to Horn, they "Segregated themselves."[85] Spatial issues tied to ethnicity were apparent in El Paso's schools in 1922 as evidenced by Horn's report.

According to the El Paso Independent School District, the school was renamed Lincoln Park School in honor of President Abraham Lincoln.[86] Bowling states, "In November 1940, it was proposed that the school change its name to Travis. The P.T.A. strongly objected, and a compromise was struck: the name was changed to Lincoln School" (no relation to the present Abraham School in El Paso's Upper Valley).[87] Clearly, the PTA, which had Mexican American families as members, had a say in the name change.

From 1916 to 1922, under Principal Demetra Stanfield, El Paso women's clubs brought school penny meals for the children. The lunches consisted of "soup with bread, beans, and milk." Bowling writes that instead of teachers giving the Lincoln students a penny for their lunches, they paid them by letting them run errands. Bowling's memoir also describes the antics of some of the children toward their teachers, including bringing beautiful flowers for their teachers from the cemetery behind the school. She also details activities surrounding the building of El Calvario Catholic Church (Filial Church of Calvary).[88] In one instance, in response to the inability of some children to bring in a quarter for school supplies, a teacher told them, "Your parents have money to give to the church, but not for your education!"[89] Economics was important to schoolteachers even then. Lincoln Park School in the 1930s and 1940s was rife with student activities and achievements, which were reported in weekly articles in the *El Paso Herald-Post* and the *El Paso Times*. Every Thursday, in the "Honor Roll in El Paso Schools," the *El Paso Herald-Post* reported student achievements, projects, and new students, such as the following highlights: "The following pupils did outstanding work in arithmetic for the second six-weeks period in Lincoln Park School: Low Fifth: Paul Grado, Rafael Hiscareno, Francisco Guerra, Jesus Manuel Lucio. High Fifth: Eleuterio Rodríguez, Ema Lara,

Carmen Garcia. Low Sixth: Juanita Telles, Lucy Imai, Jose Salcido. High Sixth: Robert Chacon, Eduardo Gomez, Santos Arciniaga. Low Seventh: Manuela Tarango, Aurora Yrigoyen, Alicia Escamilla, Alfredo Lopez. High Seventh: Tony Ortiz, Elena Loya, Natalia Zamora, Clemente Hidalgo, Willard Manley and Reymundo Puentes."[90]

Not all students mentioned in the *El Paso Herald-Post* article were Mexican Americans. The student population included Anglo students such as Willard Manley. Manley, who was born in 1921, took an interest in oratory while at Lincoln School. Willard, his sister Anne, and their brother Harry lived at 4111 Rosa Street with their stepfather Cusebio Rodríguez and their mother, Aurora. According to the 1940 US Census, Manley had dropped out of high school a year before graduating so he could work as a baker.[91]

Segregation

The creation of El Paso's early high schools followed discriminatory patterns of the El Paso Independent School District (EPISD), which redrew school boundaries to control who would attend their schools. It would not be until 1970, when the *Alvarado v. EPISD* case (which was won in 1976) challenged the district's discriminatory pattern in redrawing the boundaries, that there would be any change. Many Mexican Americans argue that EPISD's efforts to keep students from certain school areas caused long-lasting negative feelings toward schools in the district.

Historian Manuel B. Ramírez, writing on El Paso from 1920 to 1945, has argued that feeder patterns and school boundaries created segregation of Mexican students. "The Mexican schools included Alta Vista, Aoy, Alamo, Burleson, Beall, Bowie, Franklin, San Jacinto, Lincoln Park, and Zavala."[92] North Side schools included Vilas, Neill, Dudley, Morehead, Bailey, Lamar, and El Paso; East Side schools included Houston, Alta Vista, Coldwell, Crockett, Rusk, and Austin; South Side schools included Franklin, Aoy, Alamo, San Jacinto, and Bowie; Alameda District schools included Beall, Lincoln, Zavala, and Burleson, as well as the high schools Austin, Bowie, Technical Institute, Adult Homemaking, and Douglass School.[93]

Named for the abolitionist, Frederick Douglass Grammar and High School first opened in a building in Segundo Barrio in 1891 and closed in 1920. Thereafter, the school opened in 1920 at 101 Eucalyptus Street in Central El Paso. Douglass School was located less than two miles from Lincoln Park School. Lincoln Park School was considered a school in the Alameda District. At its peak in the 1930s, Lincoln School educated close to seven hundred children a year.

El Paso High School was established in 1916. Each time a school was built, EPISD board would redraw the boundaries, and school attendance was based on where students lived. Bowie High School was opened in 1927 in South El Paso in response to the overpopulation of students, but it was also due to the attendance of Mexican students in the predominantly European American and Jewish El Paso High School. Austin High School was established in 1930. Few Mexican students lived in the neighborhoods of the Austin High School feeder schools. When Mexicans began permeating the neighborhoods of the Austin High School feeder schools, EPISD built Jefferson High School in 1949.

Analysis of its curriculum and student activities shows that Lincoln Park School was a site of Americanization from its opening in 1923 to the end of school segregation in El Paso in 1955.[94] Gilbert G. González, who has written about Americanization, or the socialization and assimilation of immigrants and/or children to American ideals, states Chicano education can be divided into four periods.[95] The first period, from 1900 to 1950, represents the era of de jure segregation when Americanization took place. According to González, "During the second period 1950–65, the pattern of segregation remained, but without the deliberate official sanction of Mexican schools."[96] Mexican American culture was viewed as an impediment to the adoption of Anglo culture. The third period, from 1965 to 1975, was known as the militant and reformist era, and the fourth period spans from 1975 to present day.

González claims that historians have overlooked the Americanization of families. He states, "Educators perceived the Mexican home as a source of Mexican culture and consequently as a reinforcer of the 'Mexican educational problem.'"[97] Even though public schools were desegregated in 1955, the curriculum did not reflect it. Countless students who were profiled in the *El Paso Herald-Post* engaged in activities that did not reflect their cultural heritage. For example, an article from March 14, 1962, reported that students in teacher Miguel Franco's class at Lincoln School constructed a replica of California's Sutter's Fort.

The EPISD board desegregated schools in 1954, and in 1955 Burges High School was opened with feeder schools of Bonham, Cielo Vista, Hawkins, Hillside, and Hughey. After EPISD desegregated in 1954, students who did not live in the Cielo Vista area were given the choice of attending Jefferson or Burges High School.

When Coronado High School was built in 1962, El Paso High white students were bused to the school by the city of El Paso. When the school was built, Mexican American lawyers encouraged parents to sue EPISD to end the discriminatory practice of EPISD changing the boundary lines to benefit white students at the expense of Mexican students.

El Calvario Catholic Church

In the face of the Great Depression, Mexican Americans continued improving their community by building a Catholic Church beginning in 1932. The church acted as a satellite church of Guardian Angel Church, located at 3021 Frutas Avenue. El Calvario was located at the corner of Durazno and North Martínez Streets, across the street from Lincoln Park School. El Calvario Catholic Church reflected the life stages and ceremonies of the people from the community. Church services included baptisms, First Holy Communions, quinceañeras, weddings, and funerals. The Church of Calvary was one of several new churches that were built to accommodate the growing population. In a July 24, 1932, letter to all Catholic churches, Bishop Schuler announced "the plans for the building of two churches with El Paso, Calvary and Our Lady of Guadalupe."[98]

The Jesuits built a few of the walls of the church in 1932, but eventually Father P. Francisco Pacheco, a Claretian missionary from Spain, took over the completion of the church and built it with the assistance of parishioners. Father Pacheco, the parishioners, and volunteers from surrounding neighborhoods carried cement, sand, and rocks to the building site and up the walls until the church was completed on January 1, 1933.[99] It was blessed by the bishop three months later, on April 2.[100]

On April 30, 1933, parishioners began constructing a church tower; Father Pacheco suffered an accident when he fell 12 meters (approximately 40 feet) and dislocated his foot. The tower was completed in August 1940 at a cost of $45,000. Later, the church made other arrangements to house priests and indigents.[101] The church's seating capacity exceeded Guardian Angel Church, which was built in 1898 and is currently located a mile away from El Calvario.[102]

El Calvario's central figure was the main altar, which was comprised of three life-sized, hand-carved figures purchased from Talleres Castellanos of Barcelona, Spain, on May 25, 1933, for 20,655.00 Pts. (approximately US $146.25). It was blessed by the bishop on September 10.[103]

Activities in the church guided communal life and established the norms of the neighborhood. A former Lincoln Park resident, the late David Prieto, in his oral history interview stated that the entire community regulated its activities according to how the bells tolled. Prieto stated residents knew when someone was getting baptized, all the kids would go to the church along with the godfather. He said, "It was a big custom, and they would throw coins when they came out of church." Services held at the church included three masses on Sundays and the administration of all the sacraments. The Calvario church registry housed at Guardian Angel Church details the baptism of 5,764 children, 754 First Holy Communions, 393 marriages, and 670 funerals from 1942 to 1969.[104]

The federal government created the National Historic Preservation Act of 1966 (NHPA) and the Section 106 process, which could have preserved historical buildings like El Calvario. Unfortunately, no one told Father Pacheco or the congregation, so the Highway Department tore down the historic church in 1969 to make room for a highway pillar.[105] If Lincoln Park had been an affluent community, would the outcome have been different?

Concordia in the nineteenth century experienced dramatic changes, first with the signing of the Treaty of Guadalupe Hidalgo, which created a new international boundary, then due to the railroad's arrival. Land was platted, and El Paso's housing additions were created. El Paso saw a rise in population due to increased opportunities and the creation of transborder economics due to immigration from Mexico. Lastly, Mexican railroads brought laborers, raw materials, and cattle to the region. Concordia was molded by changing nation-states and then by border militarization after the Civil War. Hugh Stephenson and Juana María Ascarate de Stephenson settled in the Concordia tract and opened a trading post known as Stephenson's or the Stephenson Ranch. Ascarate de Stephenson established the first private church north of the Rio Grande.

For African Americans in El Paso, Mexico then became an arena of unimagined possibilities or a liminal space where in addition to race acceptance, the role of the military in peace keeping in El Paso becomes another unimagined benefit of being Black in the borderlands.

María Ascarate de Stephenson settled in the Concordia tract and opened a trading post known as Stephenson's, or the Stephenson Ranch. Ascarate de Stephenson established the first private church north of the Rio Grande. After the Civil War, Stephenson lost his land due to his sympathies with the Confederacy, but it was purchased by his son-in-law Captain Albert French, who purchased it at a fire marshal's sale and subdivided it among its heirs.

At the turn of the century, farmland dominated the landscape in the neighborhoods at the end of the city limits. The availability of water was given in the passage of the Reclamation Act on June 17, 1902, to encourage water project development, and irrigation in the western states' farmland blossomed in the region. The Reclamation Act brought much needed water to El Paso for the development of the agrarian economy. With the development of farmland, Mexican workers moved into the community. Many of these laborers moved into Segundo Barrio, Concordia, and the Val Verde communities.

Chapter 2

Race, Class, and El Paso's First Highways (1915–1958)

This chapter describes the creation of El Paso's first highway, the Bankhead Highway (or Highway 80), starting in 1915. Lincoln Park was an emerging neighborhood in El Paso's East Side at the edge of the city limits. The Great Depression in the 1930s created anti-Mexican sentiments throughout El Paso and spurred a decline of homeownership as thousands of Mexicans were deported or left the United States on their own. Redlining stigmatized these neighborhoods, and they would not recover.

The Federal-Aid Highway Act (The National Interstate and Defense Highways Act)

In her article "The Reduction of Urban Vulnerability: Revisiting 1950s American Suburbanization as Civil Defense," scholar Kathleen A. Tobin states, "Historians have given the Federal Highway Act significant credit for encouraging suburban development. . . . When the *Bulletin of Atomic Scientists* argued for 'defense through decentralization' in 1951, it gained the support of the American Road Builders Association, and lobbyists worked to persuade Congress to pass the Interstate Highway Act in 1956."[1]

The Interstate Highway System was a network of freeways that were built from coast to coast as mandated by the Highway Act of 1956.[2] A chain of other legislative events—such as the Fair Housing Act in 1949, the Urban Act of 1949, Lyndon B. Johnson's War On Poverty in 1964, and the Federal-Aid Highway Act of 1968, as well as the efforts of cities towards slum clearance—created the legal framework to build highways. The Highway Act of 1956 allocated funds to the states to purchase properties and required states to create an interstate freeway system within fifteen years or by 1972, as mandated by Congress, or they would lose the funding to build them.[3] Housing programs such as the Title VIII of the Civil Rights Act in 1968 helped to provide a mechanism for homeowners to move to other neighborhoods.[4]

The 1925 Kessler Plan (Envisioning Future Highways)

The 1925 Kessler Plan provided the foundation for city planning. The plan envisioned growth patterns and foresaw the need to create thoroughfares. Based on the recommendations of the Kessler Plan, the route for Interstate 10, known as the Thorofare Plan, was approved by Mayor Raymond L. Telles Jr. and the El Paso City Council on March 31, 1960. The report *Emerging Operations 1960: The Plan for El Paso, City Plan Commission, 2* stated, "The plan evolved from earlier studies and was updated before adoption."[5] In 1962 the City Planning Department updated the Kessler plan and created proposed highway routes throughout the city (see fig. 5). A January 8, 1962, article in the *El Paso Herald-Post* signaled the arrival of Interstate 10 with the buying of land in El Paso's East Side. The same article featured a photograph that outlined the path the freeway would take from Virginia Street to Hawkins Street and through the French and Lincoln Park communities.

When asked why Interstate 10 took the path it took, the late Manuel Aguilera, former highway district engineer, stated the decision was made in Austin because Interstate 10 was initially planned to be placed directly on top of US 80 (the Bankhead Freeway). US 80 ceased to exist once the last section of Interstate 10 had been completed.

The Great Depression, Segregation, and Redlining

In the 1930s three events had cataclysmic effects on El Paso's Mexican population: the Great Depression, the deportation of Mexicans, and redlining. Historian Yolanda Chávez Leyva explores the period of the Depression in El Paso as an important case study. She states, "El Paso's Depression-era Mexican population was heterogeneous, composed of both Mexican Americans and Mexican immigrants."[6] During these years there were some Mexican immigrants who defined themselves as temporary visitors waiting to return to Mexico and others who no longer used Mexico to define themselves.[7] A year after the stock market crash in 1929, El Paso was the home to a sizeable Mexican population. "In 1930, El Paso contained the third largest concentration of Mexicanos in the United States, surpassed only by Los Angeles and San Antonio."[8]

Chávez Leyva explains that the economic crisis in the 1930s produced massive repatriation while the National Recovery Administration (NRA) "engendered a sense of being Americans."[9] According to the Texas State Historical Association, "With the deterioration of the United States economy after 1929, between 400,000 and 500,000 Mexicans and their American-born children returned to Mexico." More than half of the Mexicans who left the United States departed from Texas.[10] Historian Manuel B. Ramírez contends that the crash of 1929 created insecurity and fear of immigrants, which in turn led to deportation and repatriation.[11] "It is estimated that one-third of Mexicans living in the United States—approximately half a million—were either deported or repatriated during the Depression," Ramírez states.[12] He argues that the atmosphere fueling deportation and repatriation of Mexican immigrants was akin to the Red Scare.[13] He writes that "the repatriation of El Paso Mexicans coincided with the deportation drive." Deportations were focused on Mexicans with illegal status and on heads of households. They were aimed at removing Mexican workers to "reduce unemployment among Anglo workers."[14]

These pressures made an already precarious economic situation even worse. Historian Ann Gabbert describes some of the economic realities Mexicans in El Paso faced during the 1930s. She states, "City-wide, eighty-two percent of Mexicans (10,260 families) rented." Gabbert wrote that the "median rent was $10-14 per month, but over four thousand families paid less than $10 per month for houses without adequate plumbing facilities or ventilation." In addition, "Slum housing for 9,500 families was concentrated on the south side below the railroad tracks, while another 500 families lived in 'equally bad conditions' in the rest of the city."[15] El Paso home and rental prices were also dependent on earnings, and Mexicans in El Paso earned truly little. Gabbert also makes it clear that people's public health was affected by where they lived. The lack of health care stigmatized poor communities; in the 1930s redlining would function similarly, branding sections of the city in ways that would affect their property values for decades to come.

The Depression in El Paso caused a disruption in homeownership as Mexicans lost their homes because of being deported to Mexico but also due to failing businesses, including the First National Bank. Communities like Lincoln Park and other Mexican areas experienced plummeting real estate values. A 1936 snapshot of the housing market in El Paso from the Mortgage Rehabilitation Division of the Home Owners' Loan Corporation (HOLC) stated, "The deportation of so many Mexicans demoralized the real estate situation in the Mexican sections of El Paso." According to the report, "Vacancies were estimated at from 30% to 50%," with "virtually no sales of Mexican property and no financing." In the report, "One real estate man said that under the present conditions there is no lending in Mexican sections 'because it is cheaper for the Mexican to move today than to pay rent or make payments on his mortgage.'"[16]

Even though the deportation of Mexicans created a difficult financing situation, the report went on to state, "The Home Owners' Loan Corporation refunded many Mexican loans," and "private enterprise readily handled the refinancing of those loans in the Mexican quarter considered to be good."[17] Historians have noted that what helped get America out of the Great Depression was World War II, but the Depression left a stigma in redlined El Paso residential areas that some would argue still exists today.

Redlining El Paso

Redlining in communities of color has played a key role in divesting minority neighborhoods like the Lincoln Park community. Redlining would have a long-lasting effect on ownership patterns in El Paso neighborhoods. Lincoln Park, as well as south El Paso, the East Side, and the future path of Interstate 10, appeared on a redlining map from the 1930s. Historian Richard Rothstein argues that federal policies created during the New Deal Era promoted segregation. Two federal agencies hampered the ability of minorities to purchase homes. The Public Works Administration program was created in 1933 to provide housing to white low-income families, and a program created by the Federal Housing Administration (FHA) in 1934 subsidized mass production of subdivisions but prohibited the sale of homes to African Americans. Housing developments received bank loans on the condition that they not sell homes to African Americans. African Americans could not buy homes in suburbs created for whites, and guidelines written in FHA manuals specified that those homes could not be resold to African Americans (which led to the creation of racial covenants). In 1926 the Supreme Court had stated that it was constitutional to create restrictive covenants against African Americans, thereby upholding racial segregation. Rothstein therefore argues that housing segregation was not merely de facto but also "*de jure*: segregation by law and public policy."[18]

In 1933 the federal government made homeownership more affordable for white families and more difficult for African American families. The process of "assessing risk" caused neighborhoods in south and east El Paso, including Lincoln Park, to be redlined by banks who would then not make loans to individuals who wanted to purchase homes in those areas or who applied for loans to remodel their homes. The national ethics code Rothstein mentions was the FHA *Underwriting Manual*, which guided appraisers, or valuators, as they were termed. Racial provisions of the 1936 FHA *Underwriting Manual* encouraged appraisers to assess neighborhoods based on racial homogeneity. Part II of the FHA *Underwriting Manual*, from a section titled "Rating of Location" with the heading of "Protection from Adverse Influences," stated, "The Valuator should investigate areas surrounding the location to determine whether or not incompatible racial and social groups are present, to the end that an intelligent prediction may be made by such groups."[19]

Valuators were instructed to verify the social and racial classes of neighborhoods. Historian Louis Lee Woods II writes that the HOLC appraisal scheme "disadvantaged low-income and minority city-Dwelling residents from obtaining financing, and by mid-century they exacerbated the disproportionately sub-standard urban housing conditions endured by non-whites in the United States."[20] Woods has stated that HOLC redlining maps and FHA underwriting manuals stigmatized minority communities that then became targets for urban renewal projects including highway construction.

All the El Paso HOLC valuators were interviewed for the *Confidential Report of a Survey in El Paso, Texas for the Mortgage Rehabilitation Division, Home Owners' Loan Corporation*, and their summaries were included in the report. One of the HOLC valuators was Ray E. Sherman, who was mayor of El Paso (1931–37) and the head of his real estate company, Leavell & Sherman.[21] Guided by the FHA, local valuators identified communities with African American populations and devalued them according to FHA policies, and they blamed Mexicans on the bleak economic conditions.[22]

Another HOLC valuator was W. R. Piper, a senior partner of Marr-Piper Agency, an El Paso management and loan firm. Speaking about "valuation shrinkage," Piper stated, "As a general proposition, values in El Paso shrunk about 45% from the 1929 level to the depression low point."[23] Based on the redlined map, Area C–designated neighborhoods were labeled as "Definitely Declining," and Area D neighborhoods were labeled as "Hazardous."[24] Based on Piper's statement, real estate values in Area D (Lincoln Park included) fell 60 percent. Adding to the mortgage crisis was the exodus of Mexicans who left their homes to return to Mexico, as well as those who were deported.

The HOLC report also mentions the deportation of Mexicans from 1933 to 1936 in a section on "Rental Advances." It stated Mexican tenement houses were purchased for investments due to requiring "little upkeep." The report also mentions that due to deportations, the rental of units by Mexicans "has become completely demoralized." In the report a common statement among real estate agents was that "it is cheaper for the Mexican to move than to pay rent."[25]

Mexican tenement houses or *presidios* usually had one central outhouse shared by all families in the same building. The upkeep of *presidios* by their owners was minimal. During this period Mexicans did not earn substantial income due to the prevailing anti-Mexican sentiment, the deportation of heads of households, and being scapegoated as taking positions from Anglos. Like African Americans, Mexicans were typically characterized as risks for home loans. The HOLC report map valued and marked communities in the following categories and colors: neighborhoods labeled "Best" had designations from A-1 to A-4 and were colored green; neighborhoods labeled as "Still Desirable" were designated from B-1 to B-6 and were colored blue; areas labeled as "Definitely Declining" were designated from C-1 to C-4 and were colored yellow; areas labeled as "Hazardous" were designated from D1 to D7 and were colored red; and neighborhoods which were designated as "Business and Industrial" had diagonal lines drawn across them and were not colored (see fig. 7).

In the HOLC map, Lincoln Park in its entirety and areas around it were colored red and labeled "Hazardous." Present-day Montana Street, two streets north of the cemetery, was considered the color line; homes located there were deemed "Still Desirable." A valuation chart assessing every neighborhood shown on the map labeled south El Paso with a D-5 designation and Lincoln Park with a D-1 designation, both "Hazardous" areas on the HOLC report:

> D-1. This is entirely a Mexican residential section wherein the houses are poor, mostly unfinished adobe.
> D-2. This small area contains the heaviest concentration of Negros in El Paso; also, Mexicans. The security therein is extremely poor.
> D-3. This large area is the concentration of Mexican peons which constitute the largest class of Mexican laborers. All the shacks therein are extremely poor and there is positively no demand of any kind for property in this section. This area, as well as all other Mexican sections, is avoided by mortgage lenders.
> D-4. This small area is also a Negro concentration point. It is also occupied by Mexican [*sic*]. There is little to distinguish adjoining D-5 except the fact that Negroes have centered therein.
> D-5. This is a Mexican tenement section. The district is occupied entirely by Mexicans and other foreigners. In this district there are many tenement houses, all cheap and in a bad state of repair.

> D-6. This strip along the Southern Pacific Railroad tracks is occupied by Mexicans, Negroes, and foreigners. The security is old and extremely poor.
> D-7. This section is similar in every respect to D-6 from the standpoint of type of houses therein and the character of the occupants.

Sections D-1 through D-7 were "redlined" and thus deemed as "Hazardous" with respect to the "security" of lending practices. In other words, individuals in the "Hazardous" areas were viewed as loan risks for bankers and creditors who followed practices created by the FHA *Underwriting Manuals*.[26]

Redlining was apparent in the Lincoln Park community labeled as D-1 in the legend of the redlined map: "This is entirely a Mexican residential section wherein the houses are very poor, mostly unfinished adobe."[27] In the 1950s properties in the Lincoln Park community were cheaper to purchase for highway building than those in the Loretto Addition, which included homes in Austin Terrace. The latter was described by HOLC in the following way: "A-4, This is Austin Terrace, the exclusive and highly restricted residential section of El Paso." In the report Austin Terrace was zoned for "high class construction" with homes "built during the last fifteen years, range in price from $10,000 to $50,000.00."[28]

As evidenced by the HOLC map, most of the neighborhoods through which Interstate 10 and Federal Highway 110 were later built, were redlined, including Latta's Woodlawn community. Ironically, the homes and businesses in the Chamizal area did not show any color markings, but they were characterized as "Definitely Declining" based on their proximity to redlined neighborhoods to the north and Mexico to the south. It is also interesting to note that land at the border that was not marked in red was farmland with dotted adobe properties that did not qualify to make the loan ranking (see fig. 7).

Racial and Class Covenants

Prior to the creation of El Paso's suburbs as a response to highway building, the African American community was landlocked in Central El Paso for several reasons, mostly economic, but also due to the existence of racial covenants and class restrictions throughout the city. For a city its size (130,000 in 1950), covenants were filed in eighty subdivisions from 1900 to 1951.[29] In addition, unplatted land in the following areas also held restrictions: all of Ysleta Grant, all the Ascarate Grant, the Fisher Survey, and the Upper Valley.[30]

Table 2.1 El Paso, Texas Subdivision Race and Class Restrictions, 1900–1951

Subdivision	Year Filed	Race Restrictions	Class Restrictions
Alameda Acres	1921	Yes	Yes
Austin Terrace	1935	Yes	Yes
AYR Lawn	1921	Yes	Yes
Collingsworth	1912	Yes	Yes
Collingsworth (Greim's)	1912	Yes	Yes
Corbin's	1922	Yes	Yes
Cotton Place	1934	Yes	Yes
Del Norte Acres	1948	Yes	Yes
Franklin Road	1940	Yes	Yes
Gateway	1924	Yes	Yes
Green Valley #1	1948	Yes	Yes
Highland Park	1903	Yes	Yes
Hillside	1917	Yes	Yes
Hughes	1932	Yes	Yes
Hughes #2	1940	Yes	Yes
Hughes (Alameda Acres)	1923	Yes	Yes
Kern Place	1914	Yes	No
Kern Place Revised	1937	Yes	No
La Sierra	1946	Yes	Yes
La Sierra Vista #2	1948	Yes	Yes
Linda Vista Gardens	1939	Yes	Yes
Logan Heights	1947	Yes	Yes
Loretto Place #1	1947	Yes	Yes
Loretto Place #2	1948	Yes	Yes
Lydia Dixon	1948	Yes	Yes

(Continues)

Table 2.1 Continued

Subdivision	**Year Filed**	**Race Restrictions**	**Class Restrictions**
McKelligon Heights Replat	1937	No	Yes
Mesa Heights	1916	Yes	Yes
Monte Vista #1	1934	No	Yes
Monte Vista #2	1934	No	Yes
Monte Vista #3	1934	No	Yes
Monte Vista #5	1936	No	Yes
Monte Vista #6	1936	No	Yes
Morningside Heights	1912	Yes	Yes
Mount Franklin View Acres	1947	Yes	Yes
Mundy Heights	1903	Yes	Yes
North Loop Gardens #1	1932	No	Yes
North Loop Gardens #2	1932	No	Yes
Pasodale	1946	Yes	Yes
Patterson	1948	Yes	Yes
Pendale Acres	1926	No	Yes
Rim Road	1928	Yes	Yes
Rosedale Farms	1923	Yes	Yes
Rosedale Farms #3	1923	Yes	Yes
Rosedale Farms #4	1933	Yes	Yes
Rosedale Farms #5	1947	Yes	Yes
San Jose	1947	Yes	Yes
San Jose Park	1947	Yes	Yes
Shanks Carpenter	1936	No	Yes
Summit Place	1915	Yes	Yes
Sunrise Acres #1	1929	Yes	No
Sunrise Acres #2	1930	Yes	No
Sunrise Acres #3	1930	Yes	No
Terry Allen Addition Block 1	1947	Yes	Yes
Terry Allen Addition Block 2	1947	Yes	Yes
Terry Allen Addition Block 5	1948	Yes	Yes
Thomas Place	1930	Yes	Yes

(Continues)

Table 2.1 Continued

Subdivision	Year Filed	Race Restrictions	Class Restrictions
Tierra Verde	1944	Yes	Yes
Tobin's Fourth	1913	No	Yes
Valley Home	1923	Yes	Yes
Valley Home Replat	1948	Yes	Yes
Clardy	1949	Yes	Yes
Cooley	1949	Yes	Yes
Loma Terrace	1949	Yes	Yes
Hueco View	1949	Yes	Yes
Moorehead	1949	Yes	Yes
Whittaker Wold	1949	Yes	Yes
Park Lane	1949	Yes	Yes
Yucca	1949	Yes	Yes
Loma Valley	1949	Yes	Yes
Trice	1949	Yes	Yes
Mesa Vista	1950	No	Yes
Rose Way Place	1949	No	Yes
Bonnie Anne Place	1949	No	Yes
Lakeside	1949	No	Yes
Miller's Lakeside	1951	No	Yes
Adobe Court	1951	No	Yes
Cielo Vista Park	1951	No	Yes
Normandy	1949	No	Yes
Sierra Vista Acres	1900	Yes	Yes
Britton Davis	1949	Yes	Yes
Unplatted land with restrictions			
All of Ysleta Grant		Yes	Yes
All of Ascarate Grant		Yes	No
Fisher Survey		Yes	No
Upper Valley		Yes	No

With permission from City of El Paso Capital Department, Assistant Director Alex Hoffman.

World War II ended the Depression. The Lincoln Park neighborhood from the 1940s to the 1950s was an idyllic but also conflicted community as detailed in oral histories with former residents. From 1940 to 1950 housing additions throughout the city appeared, but most of them included restrictive covenants that forbade African Americans from living in them unless they were servants for the homeowners. A series of historical events paved the path for desegregation in the United States and in El Paso, beginning with *Brown vs. Board of Education* in May 1954. The next year EPISD desegregated the district, but as late as 1970 there were still disparities among schools in El Paso. The state of Texas officially desegregated in 1965, and then the subsequent planning of freeways so that African Americans were able to move from the old East Side to the newly affordable suburbs that did not have racial covenants attached to them.

Restrictive covenants all shared the common clause: "No lot, site or tract of land included in these restrictions, nor any portion thereof, shall ever be rented, leased or sold or in any other manner (whether by judicial sale or otherwise) conveyed to or occupied by person or persons of Negro, Malayan, Asiatic Indian, Japanese or Chinese blood, however, this provision shall not be construed to prohibit the use of occupancy of a portion of the premises by domestic servants of the owner of the premises."[31] This meant that African Americans and other named groups could not rent, lease, or own property in those housing additions, but they surely could live there if they were working for the owners.

During the 1960s minorities were particularly impacted by highway routes throughout the United States. The late urban historian Dr. Raymond A. Mohl described the destruction that communities endured due to highway building in the postwar period: "These changes include the massive de-concentration of the central city population, the shift of economic activities to the suburban periphery, the deindustrialization or redistribution of modern manufacturing and racial turnover of the population that left many of the largest American cities with a majority black population well before the end of the twentieth century."[32] Mohl describes how the interstate system systematically have wreaked havoc on minority communities throughout the United States. Redlining aggravated both the mortgage and rental crisis created by the Great Depression and deportation in communities like Lincoln Park, East El Paso, and Segundo Barrio, which was both a Black and Mexican enclave.

Redlining in the 1930s condemned neighborhoods like Lincoln Park and others south of Lincoln Park to long-term decay by stigmatizing residents as loan security risks. "Deeming crowded neighborhoods, older properties, industrial activity, and what it called 'the presence of inharmonious racial and nationality groups' as anathema to secure investment, the FHA's *Underwriting Manual*, a veritable bible among private lending institutions, directed housing loans to the suburban periphery and opened doors for the exodus of people and capital away from urban centers."[33] The ill effects of redlining remained in El Paso neighborhoods into the 1970s. Lincoln Park's experience of a pattern of discriminatory land valuation and zoning made it easier to build several freeways through it.

Chapter 3

Community Life in the Lincoln Park Neighborhood Before the Freeways

This chapter explores the lives of Mexican Americans and African Americans and their community networks against emerging neighborhoods in the early twentieth century in El Paso's East Side located at the edge of the city limits. Like African Americans, Mexican Americans could not move into other neighborhoods because of their incomes, so they lived on the outskirts of the city or at the edge of the city limits where housing was more affordable, thus, African Americans lived alongside Mexicans.

Lincoln Park, A Black and Brown Neighborhood

Jacqueline D. Hoyt was born in Fort Worth, Texas.[1] In El Paso she attended Lincoln Park School, Sageland Elementary, and Bel Air High School in east El Paso. After high school she attended the University of Texas at El Paso (UTEP) for two years and then transferred to East Texas A&M University in Commerce, formerly East Texas State University. As a professional she worked in banking and retired from Bank of America with forty-three years of service as vice president of Transaction Services. Her family came to El Paso when she was 5 years old. Her father was an engineer with the United States

Air Force, and he was transferred to Biggs Air Force Base in El Paso. Her mother had attended Midwestern University and beauty school in Fort Worth. Her father worked as an instructor teaching pilots the instruments on planes. Her family lived at 3522 Manzana Street in Copia Street. At that time Jackie lived in a Black and Hispanic neighborhood.

Jackie said that one day, they changed her street from Manzana to Gateway Boulevard East because the freeway was coming, and their street name was going to be changed. She said that several years later, people and their friends started moving away until the time came that they too moved to a neighborhood called Sageland in the Hacienda Heights area. Jackie said she was sad because she had her heart set on attending Jefferson High School because that was where her friends were attending, and her parents told her she would be attending Bel Air (High School). She said that she did not want to go to Bel Air.

At Bel Air she tried out to be a cheerleader several times, but she was not selected. She said that at the time, the white kids at Bel Air were in most of the leadership positions. When Jackie was not selected for cheerleader, her mother went to talk to Principal Roy Chambliss about the racism at the school. Ms. Hoyt said Mr. Chambliss was bothered by her mother insinuating he was a racist and he sought Mr. R. E. L. Washington, a noted African American educator, to talk to her mother to tell her that he was not racist.

Ms. Hoyt's story documents the upheaval of Black lives due to the creation of the Interstate 10 in the late 1960s. The street named Madera, where Hoyt and her family lived, became the highway and they had to move to a new neighborhood and attend schools that were predominantly white. Such was the experience of numerous African American families in Central El Paso.

In 1959 Mateo Hinojosa Jr. said that when he returned from military service, a Jewish lady named Ms. Gould came from the East Coast to help Lincoln Park residents organize politically to vote. Hinojosa said Lincoln School was formerly a voting precinct for the area, which included the Loretto area joined with Lincoln Park residents. During that time, Hinojosa said, the Loretto area was walled off from the rest of the neighborhoods (as a gated community). Hinojosa said his mother attended a school board meeting complaining about the food the children were served at Lincoln School and questioned why the campus had not been painted. Hinojosa said the next day district painters showed up to paint the school.[2]

Hinojosa's parents bought a new house north of Raynolds Street and north of Hawkins Dairy and the Lincoln Park ponding area. His parents' house was on Ledo Road. He said they used to get flooded all the time because they had only one drain at the end of Durazno Street. "The drain was there for 50 years and whenever it rained, it caused the flooding, and the water all came from the Loretto area, and the one drain could not take it all, so it will spill over and flood the neighborhood," he said. Mateo felt the compensation for his parents' property was fair, having been appraised at $38,000. According to Mateo, highway builders advised people not to move out of the house until they received the check for their homes from the state. If they settled with the state, in turn they would "sell back" the house to residents for $200—meaning that for an additional $200 paid by the former owners, owners could go into the former property and retrieve any fixtures that were not part of the house and take them for their next home.[3]

Mateo's wife, Lydia Bueno Hinojosa lived in the Lincoln Park neighborhood from 1942 until 1964 when she married. She attended kindergarten and first grade at Lincoln School in the 1940s. Her father was in the service during World War II, and when he returned, he found out she had been getting in trouble at Lincoln School. Lydia said her father did not like the school because he felt it was not educating the children. Lydia's parents owned a store called Bueno's Grocery at 4200 White Oaks (present-day Wyoming Street), and they decided they would send their daughters to Loretto Academy instead of having them attend Lincoln School.

Mr. Hinojosa remembered the cattle stockyards in the area as well as Montes Dairy, Lanes Dairy, and Hawkins Dairy. He said Bueno Grocery had a meat section, and Lydia's father would extend credit to neighborhood patrons. He kept a ledger, and most people would pay him on Fridays when they would get paid. Lydia said there were three stores in the Lincoln Park neighborhood: her parents' store, Barrera's Grocery across the street, and Abeytia's Grocery Store located south of Hillside School. Rosa's Store (Rosa's Food Market), located at 4301 Rosa Avenue, is the only store that is still open in the neighborhood.

According to Mrs. Hinojosa, prominent families in the Lincoln Park community included "the Gutierrezes, Mr. Duran, Mr. Duron, Benavidez Family [a huge family lived next door to Concha—who was always making beans and tortillas for her large family], the Gementes, Eduardo 'Lalo' Rodríguez, City Inspector, Mauro Rosas [who was part of a family of musicians], Gerardo, Tom, Rosie, the Borregos, and a Mr. Ayala, who would train and board horses."[4] Lydia remembered Lincoln School had various grades, up to the eighth grade, with two classes for each grade. Children were severely punished for speaking Spanish. She remembered a child nicknamed Blackie who was caught defending her in the playground because she spoke Spanish. Lydia said someone told the principal and Blackie was dragged into the office and beaten. Lydia said all her teachers at Lincoln School in the 1940s were Anglos.

Every morning the children reported to their teachers to be examined for cleanliness and lice. If they were not clean, they would be taken to the bathroom and cleaned. Lydia said she remembers one boy in tenements nearby who used to bathe himself in a metal tub, in all kinds of weather, even when it snowed, so he would not be cleaned in the bathroom. Lydia's parents' house was located on what is now Highway 54.

Lydia said her parents were offered an appraisal in 1966 or 1967 and moved out in 1968. The building of US 54 did not begin until everyone had moved out. Mateo said his neighbor, Mr. Rodríguez, fought the appraisal but received less money for his home. Mateo said his parents knew the freeway was coming because his father, who was a city building inspector, had seen a letter in his office at Bassett Tower about the acquisition of properties.[5] Mateo and Lydia did not recall any community meetings that residents and business owners were invited to attend. Concerning the community life of the neighborhood, Lydia was a member of Las Teresitas, and at Easter she was part of the Passion plays; she said the entire neighborhood would participate. Lydia stated El Calvario also provided housing for indigent or homeless individuals.[6]

Mateo's sister, Martha Hinojosa Arriola, who now lives in North Corona (Norco), California, lived in Lincoln Park from 1950 to 1957. She remembers the streets in the community were not paved and flooding was a major problem. She and her sister attended Lincoln School from the second to the fifth grade. She remembers her second-grade teacher, named Mrs. Phelps. She also remembered that Ms. Lord and Mrs. Malendroff did not have good reputations (probably because they were strict). Martha was a finalist in the school's spelling bee one year. She had many happy memories of her neighbors. She recalled the Lincoln Park neighborhood as a close-knit community and that there were five to six families who used to have get-togethers. She said the Jacquez family was active with the League of United Latin American Citizens (LULAC), and the Delgado family was one of the prominent families.

Other organizations active in the community were the Parent and Teachers Association (PTA) and the Democratic Party. Martha also remembers a woman named Mrs. Ross who came to the Lincoln Park neighborhood to organize community members to vote. She said the community celebrated most festivals at Lincoln School. She learned to dance the Chapanecas and the May Day Dance. The community also celebrated all the holidays. She said the state bought their home for the creation of Interstate 10 and that fifteen to twenty families had to move elsewhere.[7]

Rebecca "Becky" Sterling said the area was racially mixed, "that there were a lot of Hispanics, a lot of African-Americans, some Puerto Ricans, and Germans."[8] When asked about how her family heard about the arrival of the freeway, Sterling said, "They contacted us (in 1958 or 1959) because they took our front yard; the city came and told us that the property belonged to the city and they took our front yard."[9] As Sterling recounts, the city of El Paso came and took their front yards to build Gateway East. It was not known whether families were compensated for the removal of their front yards.

The First El Paso Latino Texas State Representative, Mauro Rosas, Lived in Lincoln Park

Among the many students who attended Lincoln School and later became prominent persons in the community was former State Representative Mauro Rosas, a former pupil of Grace Lord, who taught there for thirty-one years beginning in 1928.[10] Rosas, the son of Justo and Julia Rosas, was born on December 5, 1925. During World War II, he served in the US Army Air Corps and later became an El Paso attorney. He was one of seven children and his family lived in Lincoln Park.[11] He attended Lincoln School in the 1930s and graduated from Bowie High School in 1943.[12] He earned his undergraduate degree in the College of Mines (now the University of Texas at El Paso) and then he studied law at South Texas College of Law. He became licensed to practice law on August 27, 1953. Rosas's two brothers, Mike and Tom Rosas, also became lawyers.[13]

Rosas's military service included the US Army, and he was part of the Army Air Corps as an aerial gunner. He was a staff sergeant assigned to the 2126 Army Air Forces Battleship and was honorably discharged after completing fifty-three combat missions in Europe. He also served as a ball-turret gunner in a B-24 bomber and his decorations included two air medals and four battle stars (ribbon with one silver battle star, good conduct medal, air medal with two bronze clusters, and one overseas service bar).[14] He was active in the following organizations: the Segura McDonald Post 5615, the Catholic War Veterans of the United States, the El Paso Chamber of Commerce, LULAC Council #8, the El Paso Bar Association, the Mexican American Bar Association, and the State of Texas Bar Association.

Rosas was also the first Latino state representative from El Paso to serve in Austin during the twentieth century in 1959 during the Fifty-Sixth and Fifty-Seventh Legislative Sessions (1959–1963). As state representative he served on the National Legislative Leaders Conference and attended the Presidential Inauguration of John F. Kennedy. He was also chair of the Free Conference Committee for the General Tax Bill, a member of the General Revenue and Taxation Committee, a member of the Constitution Amendments Committee, and served on the Criminal Jurisprudence Committee. Rosas was instrumental in creating important projects and events in El Paso. He signed the land grant to obtain land for the Sun Bowl and, with others, started the first Veteran's Parade in El Paso. He also obtained another land grant to create Veteran's Memorial Park in Northeast El Paso. He died September 10, 1993, and is buried at Fort Bliss National Cemetery.[15]

Alex (b. 1938) and Robert "Bobby" Rosas (b. 1943), Mauro's brothers, attended Lincoln School in the 1950s. During their interview Alex remembered that business owner Consuelo Forti's mother was the cook at Lincoln School. Their father, Alejandro Rosas, and his two brothers, who were born in the 1920s, also attended Lincoln School. Mauro attended Lincoln School in the 1930s. Robert attended class at Lincoln School in the 1940s during the height of segregation. As an elected state representative, Rosas advocated for Latino city employees, firefighters, bus drivers, and over time, his brothers stated, Mexican Americans were hired for those positions.[16] Alex said his brother Mauro also supported Judge José Marquez when he graduated from law school. The Rosas brothers also said there was discrimination in technical positions such as telephone pole men. Robert said Ignacio "Nacho" Padilla, former El Paso city representative, could not get hired as a pole man for the telephone company in El Paso, so he went to Los Angeles and was hired there. Upon his return he became the first Mexican American pole man hired in El Paso. He later ran for office and was elected city representative.

Alex and Robert Rosas's uncles (Mauro, Mike, and Tom Rosas) lived at their family's homestead located at 4219 Wyoming (formerly White Oaks Street). Their grandparents lived next to Ramsey Steel, a long-time business in the Lincoln Park community. During their interview Alex and Robert recounted how they felt their family was bullied into selling their property to make way for Interstate 10. Regarding the state purchase of their property for the creation of Interstate 10, Robert said, "The thing about it when they were buying the houses, the state, they were giving them almost nothing, so when they came to my grandmother's house, which would be Mauro's former home, she said, no, no, no, we're not settling until we settle for this amount of money." Roberto stated, "She was one of the few ones that got the actual value than what the state was offering at the time." He said state agents were not bilingual, and they offered owners prices below market values. "A man came in, and he was giving the owners below [market] value for their properties and since a lot of them did not understand English they were told to sign, or they put pressure on them, and they did not know they were selling their homes below market value," Rosas said.[17]

The African American Community in Lincoln Park

In the 1940s the Lincoln Park community was the eastern portion of El Paso's East Side. A sample of the 1940 US Federal Census lists five African American males and one Mexican male who worked for the railroad in the Lincoln Park community. Lee Moppins, 60, worked as a janitor in a department store and later worked for the railroad. He lived at 3204 Manzana Street. Brillo Smith, 54, worked as a head in-brakeman and lived at 308 Cebada; Robert Berry, 44, worked as a porter and lived at 3306 Manzana Street; James Frazier, 34, worked as a porter and lived at 3324 Manzana; Frank Woods, 50, worked as a boiler helper and lived at 3326 Manzana; Robert E. Lee, 46, worked as a porter and lived at 3330 Manzana; and Rafael Cruz, 34, was a boiler helper who lived at 3329 Durazno Street.[18] Later, Manzana Street was removed for the creation of Interstate 10. In 1955, the 75-year-old Moppins, who was working for the Southern Pacific Railroad, and Robert E. Lee served as El Paso delegates for the Brotherhood of Sleeping Car Porters St. Louis Southwestern Zone Conference in Dallas, Texas.

Another African American family in the Lincoln Park neighborhood was the H. L. Scales Family. Mr. Scales was born in Hotel Dieu in 1947.[19] H. L.'s father was from East Texas and his mother was from Little Rock, Arkansas. His father had come to El Paso as a worker for the Highway Department; he settled his family in the Lincoln Park community on Durazno Street in a house that he inherited from his great aunts. In an interview H. L. Scales's sister, the late Mrs. Oralee Smith stated her aunt's husband was nicknamed "Country" Claybourne and he was from Louisiana. She said her step-father was a "CCC boy" (a member of the Civilian Conservation Corps) and a cook; he arrived in El Paso in 1931 or 1932. Mrs. Smith said she knew three Buffalo Soldiers. One of them was Mr. Brown, who is buried at Fort Bliss.[20] Mrs. Smith's brother, H. L. Scales, confirmed they used to know Buffalo Soldiers who lived in the Lincoln Park community.[21]

There were other Black families in the neighborhood, such as the Jake and Annie M. Manigo family. The Manigo's home was located a few blocks from Lincoln School in the French Addition (where many African American families lived). The Manigo home was bought by the Highway Department on January 1, 1959, to create Interstate 10.[22] Ms. Smith's uncle, Mr. Harvey, bought the adobe house at 4220 Durazno in 1917. She recounted: "In 1929, a lady moved from Arkansas to Vidol [Texas], and she told Mrs. Smith's grandmother the prospects were wonderful for Blacks [in El Paso], and you could get a live-in job." Mrs. Smith said Mr. Harvey was married to Mattie Harvey, and her sister, Mamie Stevens, lived with them.[23] Mrs. Smith shared that her aunts were buried in Concordia Cemetery. Mrs. Smith said her parents lived acrimoniously, and at times she lived on Durazno Street and other times she lived on Eucalyptus Street. Her aunt, Odessa Claiborne, lived on Eucalyptus Street, near Douglass Elementary.

Mrs. Smith remembers there were not very many cars in the neighborhood, and the lions that were caged in the zoo at Washington Park would scare her.[24] She said her parents explained to her that the lions were in their cages, and they could not get her. She said without much traffic, she could hear everything in the neighborhood. She attended Douglass Elementary and Bowie High School. She graduated from high school in 1956 and attended Brown's business school until she got married. Her husband traveled throughout Europe while in the military service.

When asked about attending Bowie High School, Mrs. Smith said she and her friends organized a club called The Charms.[25] The Charms would walk to Bowie High School as a group through various neighborhoods so there would be less of a chance of being roughed up; in addition, they walked alongside the railroad tracks going to and coming from Bowie High School in Segundo Barrio. The group included Nolan Richardson and his two sisters and Mrs. Smith and her friend Dorothy, who all attended Bowie High School. Her other friends attended Jefferson High School and El Paso Technical School.

Mrs. Smith said she was inspired by Richardson, who she described as "always very calm"; her experience walking back and forth to school was terrifying, but "Sam" as they called Richardson, would put their fears to rest.[26] The Charms may have had reasons to band together for protection because neighborhoods during the 1950s were not safe for African American youth. In a January 3, 2005, article for *Newspaper Tree*,

Attorney Ray E. Rojas described the Lincoln Park community as a rough and tumble neighborhood.[27]

According to Ray Eli Rojas, the Lincoln Park neighborhood was in proximity to a neighborhood known as La Roca (the Rock) which was situated where the University Medical Psychiatric Center (formerly R. E. Thomason Hospital) now stands.[28] Mrs. Smith said African Americans started moving out of the Lincoln Park area to Hacienda Heights in El Paso's East Side. She stated that at its height (from the 1940s to the 1950s), there were more than twenty-five African American families in the Lincoln Park area, yet African Americans also lived in Segundo Barrio.[29]

In an interview the late David Prieto said in the 1950s there were five to six Black families in the Lincoln Park community. Prieto said, "What was amazing back then was that the kids, the Black kids around our age, most all of them spoke Spanish, just like us and they even spoke English with an accent, you know?"[30] He said they just saw them like regular Chicanos like themselves.

Several generations of the Prieto family attended Lincoln School from the 1940s to 1970. Francisco Prieto attended Lincoln School from 1955 to 1962. He was born and raised in the community, and his home was two blocks from Lincoln School, at 4301 Stephenson and Roosevelt Streets. He said his grandmother bought the house in 1944 while her husband was in Italy during World War II. Prieto said his family's presence in El Paso goes back to 1775, with the immigration of his ancestors from Chihuahua. His great-grandfather was also named Francisco Prieto, and he was born in 1875 in Tucson, Arizona. Prieto's grandfather, aunts, and uncles all attended Lincoln School.[31]

Concerning businesses in the neighborhood, Prieto remembered the grocery stores in the community included Lagunas Grocery, Fields Grocery, Rosa's Food Market, and Davies Market (owned by Nacho González), as well as Lane's Dairy, a Pepsi warehouse, and a trucking company. He said the Peyton Company also ran a stockyard located a block away on the west side of Lincoln School (the Union Stockyards). The stockyards were close to the railroad tracks, where they would unload the cattle and later take them to the slaughterhouse.

Prieto said his grandfather was paid $35,000 in 1970 for his house by the Texas Department of Transportation (TxDOT). Their house consisted of several lots. Prieto's uncle, David Prieto, was born in 1953 and his family lived in the Lincoln Park neighborhood.[32] His great-grandfather was born in the early teens, and he lived at 3500 Rosa. His father was born in 1909, and he lived there with his grandfather, who attended Lincoln School in the early teens. When his parents married in 1928, Val Verde Street was the end of the city limits. Prieto was born in 1953, and he was 16 years old when his family had to move from the neighborhood. He remembers that with the arrival of the freeway, the community became isolated. Prieto said in the 1930s, '40s, and '50s people fueled their homes with petroleum. He attended Lincoln School in 1958 or 1959. His older brothers attended Lincoln School in the late 1940s and the early 1950s. His father attended Lincoln School in the teens.

According to the late David Prieto, Lincoln School had three sections of every grade level, first to seventh grade, divided into A, B, and C. "So, if you were in 3A, you were in the smart class; in 3B, you were in the medium; if you were in 3C, you knew you were with the not too bright group," he said. Teachers sat you depending on your grades, with the smartest students in the first row and so on and "every student knew it was a given."[33] Activities centered on celebrating Texas heroes. Truancy was limited among the students at Lincoln School, but Prieto said he was in and out of the office mostly for speaking Spanish and for being late to class. He said he and the other children were often late because many of them went home for lunch.

Since many of the mothers did not work, at lunchtime all the Lincoln children would run home as far as four blocks away. "Many of us would be late, especially if our moms had *chile colorado con carne* or *chile relleños* or they are still making tortillas and they are still nice and warm, and we would all have a warm tortilla with butter," Prieto said. Punishment for being late to school was paddle swats in the office, and he said many of the children were late "every single day."[34]

Prieto said students at Lincoln School would also engage in antics, much to the chagrin of their teachers. He said once one student brought fresh flowers to school. The teacher was pleased when a student brought her fresh flowers until she found out he took them from a recent funeral ceremony at the cemetery.

David's sister, Dalia Prieto-Rivero, was born at home in 1957.[35] She lived at 4301 Stephenson, while Interstate 10 was built behind their home. Her parents were Francisco Prieto and Laurencia Holguin Prieto. Communities in many of the neighborhoods before the arrival of the freeways had close relationships and kinships, and they supported each other in times of need. The creation of El Paso's freeways disrupted the fabric of these communities and divided their neighborhoods.

In 1965, for unknown reasons, the Lincoln Park School was spared from demolition.[36] However, forty years later there would be a widespread public outcry against a renewed threat of demolition of Lincoln Center and indirectly against highway building in El Paso. In conclusion, even though the settlement and afterward the relocation of citizens was an unfortunate event, the narrative of the period shows that the settling of the Chamizal dispute was crucial before the building of subsequent highways such as Interstate 10, Highway 54, Highway 110, and the Chamizal Freeway could commence. If the Chamizal boundary dispute had not been settled, it would have posed an impediment to highway building because of the unsettled status of the international boundary line and of the affected neighborhoods.

El Paso's Black Wall Street

Just like the destruction of Black Wall Street, which occurred in Tulsa, Oklahoma, on May 31, 1921, El Paso suffered a similar destruction of its own Black Wall Street in 1957. Like numerous communities across the country, El Paso's African American community was not at the table when city leaders decided to deem it was blighted so they could build three freeways through it. Black residents and businesses comprised a tight-knit community before the urban changes reshaped the city's central core. Even though Blacks lived through years of spatial segregation, the community shopped, built, and attended their churches and schools and thrived in what was known as El Paso's Old East Side.

The boundaries of the city's Black Wall Street constituted the streets south of Montana Street to the north, where Texas Street meets Alameda to the east, the Chamizal before the creation of Paisano Street to the south, and the Lincoln Park neighborhood (Raynolds Street / Val Verde Street) to the east. Montana Street was the color line, or the other side of the railroad tracks, where Blacks infrequently ventured. Many African Americans also lived in Segundo Barrio.

Before the creation of El Paso's suburbs as a response to highway building, the African American community was landlocked in the Old East Side for several reasons, mostly due to economics but also due to the existence of racial covenants throughout the city. Racial covenants appeared in eighty housing additions from 1900 to 1951.[37]

The shortage of workers during World War I and the Great Migration facilitated the movement of African Americans out of the rural South, and some of them relocated to El Paso. El Paso Census records show slight increases in the African American population from 1900 to 1970: 2.9 percent in 1900, 3.7 percent in 1910, 1.7 percent in 1920, 1.9 percent in 1930, 2.6 percent in 1940, 3.2 percent in 1950, 2.15 in 1960, and 3.12 in 1970. As of July 2019, the Black or African American population in El Paso was 3.8 percent, with a decrease of 3.4 percent in 2023.[38]

In a telephone interview, Los Angeles artist Pedro Martínez, whose family was originally from San Antonio, said he grew up in El Paso's East Side around the jazz clubs with names like Club Society and The Black and Ten on Alameda Street before desegregation in the 1950s. Martínez said when he was in the ninth grade, he said he quit high school to form a rock-and-roll band, but then he was drafted and served in Vietnam. He graduated from Jefferson High School between 1964 and 1965 and played with African American bands in 1976 or 1977.[39] He remembered that there was a nightclub next to a bowling alley that featured a performance venue called Rusty's Playhouse at 3209 Alameda.

Rusty's Playhouse was indicative of the clubs, restaurants, record stores, funeral homes, and beauty and barber shops that dotted Alameda Avenue and streets south of Montana Street, which was El Paso's color line. They were part of what businessperson and director of the Eastside Central Coalition and founder of the Black Business Living Museum Michael E. P. Davis calls El Paso's Wall Street, which hosted numerous businesses opened by African Americans for their community prior to the creation of El Paso's freeways in Central El Paso. It is important to note that Black businesses coexisted alongside Mexican American businesses. Even though El Paso did not have banks like in Tulsa, Oklahoma, the economy of El Paso's Black Wall Street was a mix of Black history, Black culture, Black heritage, the Black legacy, Black religion, and coexistence alongside Mexican American culture. Many of these Black business sites are featured as part of the annual El Paso Black History Bus Tour (African American Historic Sites), which began in 2024. According to the El Paso Black History Tour, 2025 guide, "Businesses thrived during the 1940's, 50's, 60's-and 70's." They "ranged from hotels, barbershops to record stores, from funeral homes to pharmacies, doctor's offices, and music clubs."[40]

A Chicago Family in the Lincoln Park Community

The Lincoln Park neighborhood was a vibrant community. Even young children remember it vividly. Mari Primero said her parents moved to El Paso from Chicago shortly after World War II.[41] She was born in 1951 when her parents rented a house on Durazno Street. Her father fixed up the house for his family, but then the owner asked that it be returned to him once it was repaired. She remembers a flood damaged their rented home, and thereafter her family had to move to another house across from Lincoln School. In 1956 Mari's father bought a new house in the Ranchland neighborhood and her family moved away, but she remembers her life in Lincoln fondly. Even though she was 5 years old, she said she had great memories of the community. She remembered the owner of the Rosa Food Market (at 4301 Rosa Avenue) was named Nacho and his beautiful wife named Loretta or Laura. There was an African American woman named Miss Davis who also had a small store. She remembers her neighbors, the Perezes, the Cuetos (the husband was a mechanic), and a neighbor who worked for the water company. She remembers a woman named Señora (Mrs.) Gamboa, who was a good friend of her mother. Mari's family lived across from Lincoln School, next to the store at the corner owned by Juan Benavidez. The *molino* (corn mill) was located at the corner of her block, in between apartments and they shared the backyard with chicken farmers and clothes lines. Her mother's *comadre* (friend) lived in the big white house that still stands, owned by Salvador and Mercedes Leal. She remembers the Leals, who moved to California when Mr. Leal got a job for the railroad. Their son Salvador Jr. was a talented artist who graduated from Jefferson High School in 1968 but was killed in Vietnam.[42] Because his family moved to Barstow, California, he was not listed as a Texan. Mari's parents and the Leals remained friends until the Leals both passed away in their 90s.

Mari remembers the bazaars, weddings, baptisms, and the priest at El Calvario (the neighborhood church), where she was baptized and where she offered flowers the whole month of May. She remembers the trains, the funerals, the fights after school, the lions at Washington Park roaring at night, and the hobos jumping the trains. On weekends her family and their friends would all gather at Julio Cueto's house. The men would fix their cars. The kids would play. She said the women chatted and cooked. They cooked *mole, sopa*, and *menudo*. Then the men would sit around and play their guitars, sing, and laugh. She remembered that her father had a friend named Hector who did not live in the area, but he, his wife, and his family would visit every weekend. He had only one arm, but according to Mari, he was an amazing mechanic.

Leaders in the Community

Former Lincoln Park resident Teresa Orozco remembered the late Richard Telles, a member of EPISD, who helped open Lincoln School to residents so they could have neighborhood community meetings and other dedicated events. The Telles family was active politically, and the Telles machine orchestrated the election of his brother Raymond L. Telles for El Paso County Clerk in 1948, as well as for mayor of El Paso in 1957. Telles was part of the Viva Kennedy Clubs in El Paso when John F. Kennedy and Lyndon B. Johnson visited El Paso to campaign in 1960.[43]

Leased by the state of Texas, the Lincoln Cultural Center, which was the only cultural center run by the city at the time, operated from 1977 to 2006 and housed several Parks and Recreation offices, the League of Latin American

Council (LULAC); Project Amistad offices; and Project Bravo. The building featured a recreation room with pool tables, a meeting room/classroom for informal gatherings and a gallery used by local artists for monthly exhibits. The center was the only cultural arts center run by the city's PARD in a Latino/a community, so by default, we can infer it served as the city's first and to date, the only Latino/a cultural arts center. During his tenure as El Paso director of Planning, Research and Development, Nestor Valencia drew up the plans for Lincoln Park.

Although the number of Chicanos grew in El Paso from 1970 to 1980, few Mexican Americans held positions of power where they could influence local politics or the highway building that took place from 1960 to 1970. The article "The Chicanos of El Paso," a study conducted by Paul Sweeney and Carey Gelernter and published in December 1968 in the *El Paso Times*, stated "that not a single Mexican American made the top twenty-five economic 'elites' that ran the city."[44]

Politically, the mid- to late 1970s and 1980s was a period that produced organizations for the Chicano/a community. La Campaña Pro la Preservation del Barrio, a grassroots organization, was created by Segundo Barrio housing activist Carmen Felix and others in the 1970s. La Campaña focused on housing issues in Segundo Barrio. In the early 1980s, a coordinating body known as El Concilío de El Paso (the El Paso Council), composed of over twenty Mexican American organizations, was formed to advocate for Chicano rights. In Central El Paso in the 1980s, another social service organization known as Trinity Coalition, created by the Church of Christ, advocated on behalf of Chicanos in El Paso. Trinity Coalition was active in the Lincoln Park community, as were other labor justice organizations.

Humanitarian and educator Rosa Guerrero met the late Polly Harris when she was running for El Paso City Council in the mid-1970s. Guerrero said Harris referred to herself as the alder broad (instead of council member).[45] Guerrero said Harris, who had a theater background, was a big arts supporter. Guerrero said she was the most impartial person for Mexicans and African Americans, one of the few that she saw at city hall. She said it was Harris's idea to create Lincoln Center as a cultural arts center.

PART 2

Realities of Highway Building

Chapter 4

Segregation, Displacement, Demolition, and Suburbanization in El Paso (1954–1970)

This chapter touches on events from the 1950s to 1970s that affected where people lived in El Paso, which affected demolition and displacement; the creation of El Paso's freeways, which created the city's first suburbs without racial covenants; and educational segregation, which was challenged by the *Alvarado v. EPISD* case from 1970 to 1976. Urban blight, urban renewal, and gentrification are discrete and separate but are all intertwined.

The Federal-Aid Highway Act (the National Interstate and Defense Highways Act) enacted in 1956 provided millions of dollars to states and cities to construct highways as part of the mandate to create an Interstate system over a short period of time.[1] The late highway historian Raymond A. Mohl stated that the new interstates were "virtually completed over a fifteen-year period between the 1950s and early 1970s."[2] He contended that "postwar policy makers and highway builders used interstate construction to destroy low-income and especially black neighborhoods in an effort to reshape the racial landscapes of the US city."[3] This is what happened in El Paso, and in predominantly Mexican American and Black neighborhoods.

Urban renewal and slum busting were also the willing partners to create freeways. The construction of El Paso's freeways and the city's rapid growth constituted a form of social engineering and created an economic tsunami for El Pasoans. It reinforced class lines and valuated and devaluated properties, including properties that had been in families for generations. Among Mexican American residents, there was little understanding of highway building and concepts of right-of-way. In comparison, elites and those who could afford them armed themselves with lawyers.

Urban Blight and Urban Renewal Policies of the 1950s

In his book *American Urban History: An Interpretive Reader with Commentaries* Alexander B. Callow Jr. quotes sociologist Herbert J. Gans as stating "urban renewal, even armed with federal funds and the power of eminent domain, failed on many fronts."[4] Urban renewal in the United States emerged in numerous forms and laws. In El Paso urban renewal generated both blight and poverty. Actions taken by the FHA in the 1930s, resulting later in the Fair Housing Act of 1937, the Housing Act of 1949, and the Housing Act of 1954, contributed factors. In all these programs, the goal was to "revive decaying central business districts and industrial areas, preserve open space, decongest freeways, and combat air and water pollution."[5]

A publication printed in November 1957 titled *Urban Renewal for Texas* described urban renewal as "a name given to a broad range of activities in the field of urban conservation and rehabilitation for which federal loans and grants are being made available to the local community." According to the author, Clarence Schermbeck, "The program is designed to remove the causes of blight as well as to eliminate blight itself . . . [and] provides a framework within which all elements, including federal, state, and local government, private enterprise, professional and civic groups, and individual citizens can join in a coordinated effort to combat urban blight and obsolescence—in short, a program to 'help cities help themselves.'"[6]

Claims of blight by city officials assumed that there was a shared working definition of blight, whereby local governments, through their redevelopment departments, could conduct studies to identify it; cities, partnered with their respective states and the federal government, could engage in blight removal; and blighted communities, armed in their urban renewal efforts with federal legislation, could combat, reduce, deter, and eliminate blight. Citing a need for blight removal did not, on its own, automatically satisfy requirements necessary to exercise eminent domain. Politicians often took it upon themselves to designate minority neighborhoods as slums and then eradicate them.

According to historian Jon C. Teaford, "Congress launched the federal urban redevelopment program in Title I of the Housing Act of 1949, and during the next two decades, planners, mayors, journalists and the public dreamed of grand schemes to revitalize the nation's cities."[7] Teaford's opinion was that urban renewal met with limited success in the United States. He stated that with the Housing Act of 1954, "Congress required communities receiving urban renewal funds to prepare a comprehensive development plan." The guidelines were based on a plan called the "Workable Program."[8]

Even though President Eisenhower dragged his feet on certain matters, in September 1953 he appointed the Advisory Committee on Government Housing Policies and Programs to address urban issues. In his book *Urban Renewal for Texas*, Schermbeck states, "This committee was composed of representatives of major businesses, financial, and civic interests of the country."[9] In a matter of a few months, the committee produced a 377-page document titled *Report on The President's Advisory Community on Government Housing Policies and Programs, Washington, D.C. 1953*. According to Schermbeck, "The analysis and recommendations contained in the report formed the basis for the 1954 amendments to the Housing Act of Judson F. Williams, president of the El Paso Chamber of Commerce, in his presidential message to members titled "Urban Affairs Are Local Affairs," questioned the creation of the Department of Urban Affairs and Housing. He felt that the department, in company with H.R. 8429, "would create a Department of Government which could become the most powerful Cabinet post, and the most expensive."[10]

Williams, who seemed indifferent about bettering people's lives via governmental efforts, did not see the need for a cabinet-level spokesperson for public subsidy programs. He saw the creation of these programs as removing power and authority from cities, counties, and states to govern matters that were local responsibilities. When he became mayor of El Paso in 1962, he exhibited the same indifference to working on low-income housing issues in the city.

In November 1964 the Texas Highway Commission approved the largest work program in the history of the department. In an unprecedented action, the work program enabled the Texas Highway Department to purchase 3,033 miles (about the width of the United States) of land to build the Interstate Highway System and to cover construction costs for an additional 1,306 miles (about half the width of the United States).[11] As of September 1963, the state of Texas was allotted 3,031.8 miles of highways, with California following behind receiving 2,176.5 miles (about twice the distance from Florida to New York City).[12]

During widespread relocation of citizens by road construction and the lack of homes for individuals with low- to moderate-income levels, a need arose to build affordable housing. The spike in homebuilding did not occur by itself; it required federal legislation to push it forward. On June 2, 1967, Lyndon B. Johnson announced his intention to create the Committee to Rebuild America's Slums. He drew together a group of distinguished industrialists, bankers, labor leaders, and specialists in urban affairs, and asked them "to examine every possible means of establishing the institutions to encourage the development of a large- scale, efficient rehabilitation industry." The overall goal of the committee was "to develop a blueprint for the future of the American City," thereby fulfilling what would become the Model Cities Program, which was created a year after the committee was formed.[13]

President Johnson stated that the committee had a considerable influence on the Housing and Urban Development Act of 1968, and that it had identified the need for greater flexibility in the housing market by providing federal subsidies (tax credit funding) for low- to moderate-income housing.[14] The 1968 Housing and Urban Development Act guaranteed financing for private entrepreneurs to plan and develop new communities. The committee also recommended making participation in federal programs more attractive to private companies and proposed the creation of the National Housing Partnership to involve private-sector companies in building housing for the poor.

In the 1960s, with the efforts toward decentralization of city centers, land speculation was driven by special interests more than community needs, and neighborhoods were often caught in the middle.[15] This period also coincided with desegregation and the arrival of African American enlistees and officers who wanted to live in suburbs, but due to existing racial covenants, they could only live in certain neighborhoods or in new neighborhoods that did not have those restrictions.

From 1967 to 1970 new additions were created by the city to house the shift of population from Central El Paso, but home building in Lincoln Park in 1970 remained static. Tobin has stated, "Though suburbanization in the United States during the 1950s is a well-known story, scholars still consider postwar prosperity and underlying desire on the part of the American people to move further away from the problems of the inner city as its primary causes." The thesis in her article states decentralization was due more to fear of the atomic bomb and being attacked by foreign powers, and the perception that moving into suburban areas reduced "urban vulnerability."[16] Decentralization of cities shifted populations from downtown areas to city outskirts and suburbs. *The Historical Overview of Housing* issued by the Department of Planning Research and Development stated, "The most striking evident trend during the period 1940–1950 was the movement towards single-family, detailed dwelling units and away from multi-unit structures."[17]

In the *1950 Housing and Population* study, data from "Persons of Spanish surname" was collected for the first time in the US Census. Before the relocation of residents due to highway building there was already a shortage of housing in El Paso. The *1950 Housing and Population* data for the El Paso Standard Statistical Metropolitan Area (SMSA) in El Paso County shows 45.9% of the population had Spanish surnames, but they occupied only 37.5% of the dwelling units.[18] About 45.1% of total residents were homeowners, but only 29.4% of those with Spanish surnames owned their homes, and 36.5% of those with Spanish surnames lived in dilapidated housing compared to 18.4% of the total population.[19]

In its 1964 publication, *Historical Overview of Housing in El Paso*, the City of El Paso's Planning Department estimated that in addition to the loss of commercial and government structures, the Chamizal Settlement with Mexico would result in a loss of 1,253 dwelling units occupied by 1,155 families.[20] The reality proved close to that estimate: between 1963 and 1966, 741 tenement units were demolished in fulfillment of the 1964 Settlement, and 355 owner-occupied, single-family dwellings in Cordova Island were purchased by the International Boundary and Water Commission to be transferred to Mexico.[21]

In 1954 the Amendment to the Urban Act via the Housing Act sought to rehabilitate existing structures, curb the demolition of structures, and solve issues of blight. In 1964 the War on Poverty created opportunities to fix housing issues with programs such as the Workable Program, a stop-gap effort to help homeowners displaced by signing the Chamizal Settlement.

At the local level, El Paso Chamber of Commerce leaders questioned whether a federal department of urban affairs would provide any support for local municipalities with the creation of President Johnson's Committee to Rebuild America's Slums, which later became known as the Model Cities Program. Johnson followed with the creation of the Housing and Urban Development Act of 1968. The 1960s also coincided with the Civil Rights Movement. Urban historian David R. Díaz states "the Civil Rights movement made two significant social demands: political and civil rights and economic empowerment."[22] Part of economic power was the right to affordable housing.

El Paso benefitted by the creation of organizations involved in President Johnson's War on Poverty. Programs such as the Workable Program, created through the Economic Opportunity Act of 1965, is explored in this chapter. Under the Workable Program local grassroots groups like MACHO (Mexican American Committee on Honor Opportunity) and Project BRAVO (Building Resources and Vocational Opportunities) were born. The quest for affordable housing resulted in the demolition of hundreds of housing units in the Chamizal community, which were deemed blighted. This chapter shows that the settlement of the Chamizal in 1963 spurred widespread relocation of residents, many of whom did not receive comparable funds for their homes since they were classified as blighted and inferior.

El Paso County's loss of acreage and simultaneous rapid population increase had a profound impact on housing, as did *Brown v. Board of*

Education of Topeka, 347 US 483 (1952–1954) and later the *Alvarado v. EPISD* (1970–1976) case that forced desegregation in El Paso schools. The city's population grew exponentially in the later part of the twentieth century, which created new opportunities and dilemmas. Escalating needs for housing and infrastructure in the growing metropolitan center supported the case to build freeways, while city leaders used school boundary lines to limit Black and Mexican attendance in white schools.

Slum Clearance

The Chamizal Settlement and the building of Interstate 10, Highway 110, US 54, and the Border Highway were all significant factors in the creation of El Paso's suburbs. During his first term, from 1951 to 1955, Mayor Fred Hervey, a proponent of slum clearance, gave the city of El Paso carte blanche to condemn private properties and begin reshaping the city in preparation for highway building and real estate development. An article in the May 12, 1954, issue of the *El Paso Times* by Peter Schrag detailed citizen discussions of the new city charter that granted new powers to the city. The charter included a section on urban development, or "urban renewal," which gave eminent domain authority to the city. According to Schrag, the charter provided "the power to condemn private property which is substandard or blighted and resell it to another private owner who indicates that he will use the land or property in such a way as to conform with improvements outlined by the City and Chapter 2 of the city charter allowed the city to accept State and Federal aid in slum clearance and development work."[23]

The Kessler Plan had envisioned the creation of thoroughfares in El Paso as early as 1925. However, because properties in the Lincoln Park community had been redlined as early as 1936, homes there were valued at significantly lower prices than homes a few blocks away to the north. Was it social engineering or just the opportunity to make money? It was not illegal, but developers used insider information to capitalize on the opportunity to build dense, low-cost housing in what had been desert and sand hills in the mid-1950s. Was it strictly business? Developers used insider understanding of highway plans and the need for inexpensive tract homes. Developers positioned themselves to build subdivisions west of Bassett Center in the new east El Paso (south of Interstate 10), in areas annexed by the city of El Paso on March 15, 1955.[24] These subdivisions included Ranchland Hills, Del Norte, Hacienda Heights, Cedar Grove, and others.[25]

The creation of Interstate 10 established a class line across the city (not unlike the cliché notion of neighborhoods distinguished by which side of railroad tracks they occupy). The class-based economic line meant homes built east of Hawkins Boulevard and south of I-10 were inexpensive compared to those built north of Interstate 10. Homes and businesses located north of the interstate appraised at a higher value than homes located to the freeway's south. Homes south of the freeway sold for under $10,000, while homes north of I-10 sold for over $10,000.

The Federal Housing Act of 1937 triggered the initiation of slum clearances in Southwestern cities like Tucson.[26] According to Sarah F. Liebschutz, three laws helped to address problems in American cities in the late nineteenth and early twentieth centuries. The first was the National Housing Act of 1949, which "was established for widespread and continuing national support for housing and community development activities." According to Liebschutz, the subsequent Housing Act of 1954 "modified the emphasis on urban renewal by encouraging rehabilitation rather than demolition of existing structures, and stipulating participation of neighborhood residents in the planning process."[27]

The creation of freeways also contributed to the decentralization of cities as people moved away from densely populated areas. The concept of urban renewal, and how cities participated in programs to act as proponents of it, was closely tied to housing issues. It resulted in a chain of legislation, including the Fair Housing Act in 1949, the Interstate Highway Act in 1956, and others focused on slum clearance, like the Urban Act of 1949 and the War on Poverty in 1964.

An article in the May 12, 1954, issue of the *El Paso Times* detailed a city plan to condemn private property deemed substandard or blighted.[28] The condemnation of private properties by the city coincidentally occurred at the same time as the 1954 amendment to the Housing Act of 1949.[29] We have since learned that residents who lived in the neighborhoods that were considered blighted did not think of their communities in the same way, but since residents in those areas did not possess agency, they did not have the tools to fight against those designations.

The Fair Housing Act declared every American deserved a "decent home and a suitable environment." Robert E. Land and Rebecca R. Sohmer, writing about the legacy of the 1949 Housing Act, stated that Congress attempted to achieve this goal with various programs: "Title I financed slum clearance; Title II increased authorization for Federal Housing Administration mortgage insurance; Title III committed the federal government to building 810,000 new public housing units; and Title V allowed the Farmers Home Administration to grant mortgages to encourage the purchase or repair of rural single-family homes."[30] These forces created a unique opportunity for cities to make slum clearance and highway development possible.

A 1956 map of El Paso shows the Lincoln Park community at the edge of the city limits. Not all homes were connected to the city's sewer lines, which meant that parts of it could be defined as slums, rendering property values lower than those north of Bassett Center. Mateo Hinojosa's parents' home on Ledo Street is an example of a home that was declared to be in a slum or blighted area. His house, along with all the houses on his street, were purchased at under-market prices to construct Interstate 10.[31]

The Workable Program, Project BRAVO, and MACHOS

Working with grassroots organizations, the federal government created the Workable Program in response to the Economic Opportunity Act of 1965, which "called for the maximum feasible support of the poor."[32] Under the program communities were required to participate in urban renewal efforts. In El Paso grassroots groups like the MACHOS and Project BRAVO participated.[33]

The 1962 program guide for the *Workable Program for Community Improvement* covered the following: 1) Codes and Ordinances; 2) Administration Organization; 3) Comprehensive Community Plan; 4) Financing; 5) Neighborhood Analysis; 6) Housing for Displaced Families, and 7) Citizen Participation.[34] A goal of the program was to foster community participation and lay the groundwork for neighborhood support of efforts to combat blight.[35]

The Alameda Policy Advisory Committee of Project BRAVO collected data and developed program plans, which it presented to the El Paso City Planning Department in its application for a Community Renewal Program. Project BRAVO and other organizations, such as MACHO, also assisted the Planning Department in identifying "blight" in their communities.[36]

The Workable Program was geared to assisting displaced families. The creators of Project BRAVO present a counternarrative to the boosterism of the period, contradicting the rhetoric of the El Paso Chamber of Commerce and other groups that promoted mass demolition of homes deemed as "blighted."

In 1964 the program provided approved right-of-way funds as part of a city-sponsored bonds issuance and an Urban Beautification Program of half a million dollars a year to counter urban decay in declining and low-income areas.[37] Another objective of the Workable Program resulting from the Chamizal Settlement was border improvement. To help Chamizal residents with relocation, Public Law 88-300, S. 2394, known as the American-Mexican Chamizal Convention Act of 1964, was passed by the Eighty-Eight Congress, on April 29, 1964.

According to the *El Paso Development Manual, Urban Development Manual for the City of El Paso: A Handbook for Community Planning*, "The area just north of Cordova Island (from Paisano to Alameda, and from Piedras to the North-South Freeway, comprising 153 acres) was classified in the 1960 census as about 52% blighted."[38]

In contrast to Mayor Hervey's first administration, in his second administration a 1975 Community Development Program, financed from the Housing and Community Development Act of 1974, outlined the need for funding to keep Lincoln Park, and Lincoln Center, from becoming slums or blighted areas. The actions that were assigned maximum feasible priority for the prevention of slums and blight and for improvement of the environment in the proposed first-year budget among the three-year programs seeking community development block grants for 1975, 1976, and 1977 included renovation funding for the Lincoln School.[39] Lincoln School was renovated to create a cultural fine arts center, and the open space that had once been residential space around the school became Lincoln Park.

LBJ's War on Poverty and Project BRAVO

The creation of Lyndon B. Johnson's War on Poverty in August 1964 provided job training for the poor and marginalized. That was the same year that the Chamizal Settlement was ratified, causing community members to ask themselves how they could help poor citizens. In El Paso a community action organization, the El Paso Community in Action Program, Project BRAVO, was created under provisions of the Federal Economic Opportunity Act of 1964, and it became a beneficiary of Johnson's War on Poverty.[40] The organization sought to respond to economic conditions faced by Mexican Americans in El Paso in the 1960s. In September of 1964, Salvador "Sal" Ramírez, associate director of the Boy's Club of El Paso, met with social workers to discuss the War on Poverty, or the Poverty Bill, as it was termed.[41] Ramírez was familiar with many members of El Paso's Jewish community, who were also members of various agencies concerned about poverty in El Paso. The individuals who came together in 1964 did so because they sensed an urgent need for action in the wake of the reorganization of the US- Mexico border following the Chamizal Settlement.

A document in a spiral-bound binder, in a box labeled "Upward Bound & Project Bravo," belonging to Dr. Ralph Segalman (perhaps his personal copy, as the cover bears his signature), recounts the history of the creation of Project BRAVO. At the time Segalman was a doctoral social research candidate at the New York University Center for Human Relations and Community Studies, engaged in field research with the Jewish Community Council and Center of El Paso.[42] Project BRAVO was not the subject of his dissertation.

Several individuals associated with the creation of Project BRAVO were members of El Paso's Jewish community, and they loaned Segalman to Project BRAVO on a part-time basis. The spiral-bound document offers a richly detailed account of the counterstory of the Chamizal Settlement and the need for an organization like Project BRAVO. Unlike other reports from the city's Planning Department, it does not wallow in the boosterism of the period.

Segalman's report states, "The transfer to Mexico of the Chamizal, which was largely a slum and industrial area, and the relocation of its residents to housing they can afford are bound to create a chain of social strains spreading throughout the city."[43] Here for the first time, a group of informed citizens presented a counternarrative concerning relocations. Jewish leaders and members of El Paso's social agencies reasoned that the people relocated from the Chamizal would face financial hardships that could affect the entire city.[44]

The report's findings were in direct disagreement with the rhetoric of International Water and Boundary Rights Commissioner Joel Friedkin, who, along with city planners and the El Paso Chamber of Commerce, claimed that the changes mandated by the Chamizal Treaty were needed to bolster the economy. In contrast, those associated with Project BRAVO did not see the Chamizal Treaty as a win-win for the entire community and were joined in their objections by other socially conscious organizations. The Texas Employment Agency, the Sociology Department of Texas Western College, the El Paso Boy's Club, the El Paso Chapter of the Texas Social Welfare Association, the General Practice of Medicine, Catholic Counseling Services, the Catholic Youth Organization, the Mathematics Department of Texas Western College (present-day University of Texas at El Paso), along with several independent citizens, were allied with Project BRAVO in expressing their concerns.[45]

Members of the Jewish Community Council and Goodwill Industries became staff members of Project BRAVO.[46] Project BRAVO had a long-lasting, positive impact on El Paso, and paved the way for later community programs and action groups in the South Central and Lincoln Park communities, such as the Weed and Seed Program, which was created in the 1970s to curb gang activity in the area.

The need to settle the Chamizal dispute was revealed to be the reason for creating the Community in Action Program, to address pressing needs for employment, education, neighborhood development, and health. Those needs, which resulted in the creation of Project BRAVO, are reflected in Segalman's report: "El Paso is confronted by serious, complex, and terrible problems created by its rapid growth and by its large and expanding slum areas and poverty-stricken populations."[47]

Segalman's researched included this information on poverty: "Twenty-five percent of the city's population lives in slum areas, and this is aggravated by the fact that over 11 percent of El Paso families earn less than $2,000 annually." The report also included that the 1960 census showed 3,023 families with less than an income of $1,000, 4,859 families with an income of less than $1,999, and 15,358 families with less than $3,000.[48] Not only were many families in the Chamizal of modest means, but they would also be displaced by the signing of the 1964 Chamizal Settlement.

Chapter 5

International, National, State, and Local Issues on Highway Building

While key actors at federal, state, and local levels were shaping El Paso's urban space through the creation of its freeways, various forces were at play such as desegregation and due to highway building the planning of the city's first suburbs. Unfortunately, the communities affected by these changes, who would be moved and displaced, were not at the table. At the local level, the decisions for the creation of El Paso's freeways were at the hands of representatives from the Highway Department, which was a committee under the chamber of commerce. One important figure in the creation of Interstate 10 and subsequent roads was attorney Tom Diamond. Diamond created the city's first right-of-way office, which initiated the process of acquiring land for the Texas Highway Department. Diamond later served as chairperson of the El Paso Chamber of Commerce's Highway Council after the passage of the Highway Act was federally mandated in 1956. Another important individual was El Paso District Engineer Joe Battle, whose tenure at the Texas Highway Department from 1963 to 1991 coincided with the building of El Paso's first freeways and its major roads (for more information about these individuals, see the appendix).

Two of El Paso's freeways, Interstate 10 (built in 1967 and completed in 1968) and Interstate Highway 110 (built in 1971) were built with state, city, and federal funds. The North-South Freeway, or Highway 54 (built in 1968), and the Chamizal Border Highway (built in 1972) were built with federal funds. Properties for the creation of Interstate 10 and Highway 110 were acquired via the Texas Highway Commission from 1964 to 1970. Properties for Highway 54 were acquired by the federal government, and the Chamizal Freeway (Border Highway) was created as a condition of the Chamizal Settlement, which was signed in 1963 (ratified in 1964). Note that highway building in El Paso did not begin until after the Chamizal dispute was settled. Interstate 10 ran parallel to the Bankhead Freeway (Highway 80), which ran parallel to the railroad tracks and the Rio Grande.

The 1964 Chamizal Settlement's Influence in Building El Paso's Freeways

The placement of El Paso's first highways also centered on decisions on how the city was planned in context to its transborder location. Highway building was also dependent on the rectification of the Rio Grande, which according to the Treaty of Guadalupe Hidalgo of 1848, was the boundary line between the two nation states. In El Paso the boundary included an area known as the Chamizal.

The Rio Grande had changed course several times over the years—in 1827, in 1852, and finally in 1880 when it created Cordova Island, which was considered Mexican territory.[1] The signing of the American-Mexican Chamizal Convention Act of 1964 displaced residents all along the US-Mexico border as parts of El Paso were returned to Mexico and other parts of Mexico were returned to the United States. The portion that was not returned to Mexico was turned into the Chamizal National Memorial and into Bowie High School. The Chamizal Settlement became the test case that then sanctioned the relocation of residents to other neighborhoods, like the Lower Valley and northeast El Paso.

The Chamizal Settlement was signed in 1964 and solved a centuries-old issue on each side of the US-Mexico border. Since rental and private properties in the Chamizal were lower than in other parts of the city, individuals who were relocated either moved to low-cost housing subdivisions, housing projects in east El Paso and the Lower Valley, or neighborhoods like the Devil's Triangle in northeast El Paso. In other words, relocation was based on what citizens could afford and on how much they received for their properties, which was not substantial because they had been redlined or had depreciated over time.

In his oral history interview, the late Tom Diamond stated the I-10 route was dependent on the cost of the acquisition of the land. He said a Texas Highway surveyor said, "The location of the freeway had been determined by the North line of the Ysleta and the Socorro grants."[2] According to Diamond, the I-10 site selection was based on the land grant boundary lines in effect at the time. I-10 changed the east-west subdivision of the French and Lincoln Park communities into north and south halves with the boundary being I-10.

Additionally, the Chamizal Settlement was a turning point and major driving force in the redevelopment of Lincoln through road transportation projects. Specifically, the treaty mandated the building of the Chamizal Highway (later renamed the Border highway) and the Port of Entry Cordova Bridge (a.k.a. the Free Bridge). The site selection of the port of entry aligned to the north with Highway 54, and as a logical result it was connected to I-10 via I-110 and Highway 54, which later became the North-South Freeway. The result was a division of the French and Lincoln Park communities into four quadrants.

The area of the Chamizal itself was what historian Jeffrey M. Schulze described as "contested boundaries between colonial domains." Due to the unpredictably changing nature of the river (which was the international boundary between Mexican and United States), the boundary would change whenever the river crested due to rains and flooding. The area "attracted thousands of residents who often quickly recognized and capitalized upon the tract's uncertain status."[3] As of the mid-1960s, 3,750 residents lived there.[4]

On February 29, 1968, F. C. Turner, director of Public Roads for the US Department of Transportation, sent a letter to J. C. Dingwall, state highway engineer for the Texas Highway Department, notifying him that he could proceed with building the Chamizal Border Highway. The letter authorized a budget of $3 million dollars to build the freeway.[5]

The Chamizal Settlement and the Chamizal Memorial Freeway factor into examination of the Lincoln Park neighborhood because they were barometers of Interstate 10 and the North-South Freeway. To build the Chamizal Freeway, the state of Texas took properties from two blocks west of Santa Fe Street to Zaragosa Road.[6]

The vast acquisition of properties was a catastrophic event in El Paso–Ciudad Juárez history. It is also interesting to note that former Mayor Fred Hervey (1951–1955) was chairperson and arbitrator for the Chamizal Planning Commission. He would later be reelected mayor of El Paso for a second term, from 1973 to 1975, and he and his brother would build their real estate empire.

El Pasoan Michael Patino's parents, who were renters in the Chamizal neighborhood, received a series of letters in 1970 from District Engineer Joe M. Battle of the Texas Highway Department about the acquisition of their property.[7] The Texas Highway Department hired relocation officers to assist citizens whose residences were acquired.

The purchase of properties and the relocation of people were tests for the massive, forced relocation of residents in later years to build Highway 110, the North-South Freeway, and the link to the BOTA. While the totality of the Chamizal Settlement is beyond the scope of this study, it is important in the context of the building of the Chamizal Freeway, which also displaced citizens and changed their lives. In Tom Diamond's description of right-of-way property acquisition practices, there were no negotiations; individuals had to accept what they were offered and were required to move.[8] Relocation of Chamizal residents resulted from two events: one, the return and exchange of lands—Cordova Island returned to Mexico and the swap of the Rio Linda neighborhood—and the other, the creation of the Chamizal Memorial Freeway, which later became known as the Border Highway. Due to its proximity to the US-Mexico border, the Chamizal Highway was a federal project managed by the International Water and Boundary Rights Commission. The Chamizal Highway took properties from the Campbell Addition south of Segundo Barrio and the Cotton Addition.

The Chamizal was a conglomeration of neighborhoods and subdivisions that included Cordova Island. City planning documents state that a sizeable number of homes lacked utilities, had outhouses, and required sewer systems. Scholar Maria Trillo, in her ethnography of the Chamizal community, writes a very different description of the Chamizal: "El Chamizal contained four principal components with it: two communities, an industrial/warehouse area, a large cotton field with the remains of a once-thriving cotton gin named the El Paso Milling Company, and a radio station (KSET) in the middle of those cotton fields."[9]

The Rio Linda Neighborhood

According to Trillo, "One of the communities was Rio Linda, and the other was known as Los Jardines [Los Bosques] de Córdoba/Cordoba Gardens or Los Diablos (The Devils). The area also included an industrial area had several warehouses and industries, including the Peyton (Meat) Packing Company, Three-V Cola, and a bleach factory." The area was part industrial, part rural, and had "private-cotton fields owned by non-Chamizal residents like Texas College of the Mines" (now UTEP).[10] Trillo states that the area also included cattle corrals, the Rio Grande, and a "western/cowboy" movie set that was used by Hollywood from time to time. The US-Mexico boundary was located at Cordova Island, also known as La Isla de Córdova. La Isla had a US international bridge and immigration/customs that linked Mexico and the US and was part of the Chamizal dispute (see fig. 2).

Trillo's neighborhood of Rio Linda was considered a middle-class enclave (homes had utilities and a sewer system) by Segundo Barrio residents. It consisted of single-family brick structures. As part of the Chamizal Settlement, the Rio Linda neighborhood transferred to Mexico in the context of the land swap. Because Mexico's sewer system was incompatible with the Rio Linda houses, the fifty homes transferred were demolished.

William E. Wood Jr., a former government appraiser during the Chamizal settlement, said Chamizal was a large area and not just a neighborhood. Residents were offered their property values based on what their homes were worth. Wood stated that most of the homes were humble and were built after World War II. In his interview for the Chamizal Oral History Project, he stated, "They were little houses in the 900 [and] 800 square foot range, most of them, some of them would be around 1,200 to 1,500 square feet)." He stated that the "houses were immaculate [and] well-kept, and that pride of ownership was extremely evident."[11]

Wood was one of two appraisers during the relocation of residents. Moving expenses for the relocation of people were paid by the government. In 1967 the Texas Highway Commission approved the Chamizal Highway as part of the November 8, 1964, treaty. The 12.5 miles of the Chamizal Freeway starts two blocks west of Santa Fe Street, passes through the Chamizal area, and ends at the Zaragosa Bridge. Congress authorized $8 million in federal funds to pay for it.[12] The amount increased to $15 million in 1967.[13]

The *El Paso Chamber of Commerce Magazine* reported, "With assistance from the Chamber's Government Relations Committee, headed by Chairperson Mart Pederson, we have helped make a reality out of the Chamizal Memorial Freeway, approved by a joint Texas Highway Department [and] US Department of Transportation project." When asked how many automobiles would use the Chamizal Memorial Freeway, Highway Committee Member Tom Diamond stated, "The US-Mexico boundary from downtown El Paso to the Zaragoza port of entry will provide a vital El Paso link estimated by the State Highway Department to carry over 40,000 vehicles a day."[14]

In just four years, the Chamizal dispute went from being an opportunity to a problem, according to the rhetoric in the El Paso Chamber of Commerce's publication *El Paso Today*.[15] In an August 1, 1962, issue, an article titled "El Pasoans Support Chamizal Settlement Plan" was featured.[16] The rhetoric in the article favored the plan, while according to late historian Leon Metz, dozens of letters against the settlement were sent to the editors of both the *El Paso Times* and the *El Paso Herald-Post*. Both newspapers countered by publishing editorials in favor of the agreement.[17]

Amid the boosterism that surrounded the proposed settlement of the Chamizal, one person who was pivotal to opening the community dialogue against the project was Elvira Villa Lacarra Escajeda, profiled in a 2006 short film titled *Vila* by Arturo and Vallarie Enriquez, from Vantage Point Studios in El Paso. In the video Vila, who later became the founder of the Chamizal Civic Association, speaks of reading about the Chamizal issue in the newspaper and then going to the Paso Del Norte Hotel to speak with

US Ambassador to Mexico Thomas C. Mann during his visit to El Paso.[18] In the video Vila states that she waited hours to see Mann, until someone came out and told her it would not be possible for Mann to meet with her. She then called the *El Paso Herald-Post*, and Marshall Hail arrived at the hotel to interview her.

When Hail arrived Mann came out of his meeting and apologized to Vila, and told her, "I apologize, but your Mayor had told me that everyone was 100% for the Treaty."[19] Hail reported that Ambassador Mann "assured the 'little people' of south El Paso he would get their views of a proposed Chamizal Zone settlement before their property is transferred to Mexico."[20] On Monday, April 13, 1963, US Representative Ed Foreman of Odessa met with nearly eight hundred Chamizal residents and business owners at Sacred Heart Church and took questions about the issue.[21]

Federal and state departments lacked bilingual relocation agents, which necessitated other methods to educate Spanish speakers about the concept of Right-of-Way. A 15-page Spanish booklet titled *La Compra Del Derecho De Via* (*The Purchase of Right-of-Way*) was printed by the El Paso City Planning Department in conjunction with the Texas Highway Department (see fig. 6).[22]

In the introductory comments, the booklet stated that the publication's purpose was to answer questions about the state's right-of-way practices. The booklet detailed which negotiations for the purchase of properties for right-of-way were handled directly by the State Highway Department, which was responsible for the acquisition of properties. Since freeways were new to Spanish speakers, the Highway Department expounded such concepts as "The engineers design and place the freeway with you in mind, to give you service and security and to reduce the price of operating your vehicle."[23]

To encourage the relocation of residents, the government paid closing costs for the purchase of relocation homes for Chamizal residents.[24] The last page of the booklet explained the process of eminent domain ("Proceso de 'Dominio Eminente'") for individuals who refused to sell to the state.[25] Historian David Dorado Romo has stated that the ratio of people relocated in the Chamizal dispute was six Mexican Americans to one Anglo resident.[26]

National Perspectives on Highway Building

Highway building in the United States and in Texas was an orchestrated series of decisions made by federal, state, and local politicians, lobbying groups such as the American Association of State and Highway Officials (AASHO), and other interested parties, as well as chamber of commerce groups and business elites.[27] In Texas before 1917 the state did not have a transportation commission, and road building was left in the hands of counties (and from 1870 to 1890, of local municipalities).[28] AASHO's creation coincided with the development of the first intercontinental highway in the United States, known as the Bankhead or Highway 80, which was constructed in 1916. As evidenced in other Southern states, from 1890 to 1917, road building, as in the creation of the Bankhead Highway, was built via convict labor.[29] Convict labor during this period was seen as a form of reforming or rehabilitation for prisoners.

What cannot be overlooked in the creation of freeways and suburbs was the role of the automobile, "which demanded hard-surfaced roads." Urban historian Raymond A. Mohl states that a popular General Motors Futurama exhibit at the 1939 New York World's Fair, according to Mark I. Foster, "stimulated public thinking in favor of massive urban freeway building." The designer of the Futurama exhibit "also promoted the idea of a 'national motorways system connecting all cities with populations of more than one hundred thousand." President Franklin D. Roosevelt met with Thomas MacDonald in 1938 and tasked him to create a report on the idea of creating "a system of east-west, north-south transcontinental highways."[30] MacDonald produced his report, entitled *Toll Roads and Free Roads*, in 1939. Another report, entitled *Interregional Highways*, was created in 1944 by the National Interregional Highway Committee, again headed by Thomas MacDonald, who was appointed by President Roosevelt. It promoted the idea that "dislocated urban residents . . . could move to the new suburbs and commute to city jobs on new high-speed expressways."[31]

In September 1955, a year before the passage of the Highway Act, the US Department of Commerce, Bureau of Public Roads, under Dwight D. Eisenhower's administration, issued *General Location of National System of Interstate Highways, Including All Additional Routes at Urban Areas, Designated in September 1955*.[32] Known as the "Yellow Book," it included three sentences of text and maps for 43 states and the District of Columbia where the Interstates would be built.[33] The Yellow Book designated the future routes of highways. It was referred to as the Yellow Book because of the color of its cover.[34]

The Cold War's Influence on Highway Building

From the mid-1950s and early 1960s, highway building and subsequent suburbanization were linked as issues of national security. Scholars such as Kathleen A. Tobin have stated that Cold War historians saw "the link between suburbanization and atomic fears."[35] People sought to move away from downtown areas to distance themselves from the city center in case of an atomic attack. In Charles B. Quattlebaum's view, writing about highways in 1944, the first American roads were military, and the army was the principal instrumentality of road construction in early American history. According to Quattlebaum, "military roads were built on Native American trails, and later settlers created them during the conquest of the Northwest" in the late eighteenth century.[36] "When the *Bulletin of Atomic Scientists* argued for 'defense through decentralization' in 1951, it gained the support of the American Road Builders Association, and lobbyists worked to persuade Congress to pass the Interstate Highway Act in 1956," Tobin wrote.[37]

Working hand in hand with city planning departments and state transportation agencies, federal agencies worked to clear the land to build highways. Because highway building was linked to issues of national security in the Cold War era, transportation maps in the mid-1960s were presented as Defense and Highway Maps at public hearings.[38] From October 30 to November 10, 1961, El Paso hosted a two-week National Security Seminar planned and programmed by the Armed Forces Industrial College and sponsored jointly by the El Paso Chamber of Commerce and Texas Western College. According to Chairman Carl W. Connors, general manager of the Mountain States Telephone and Telegraph Company (El Paso), "The objective of the seminar is to bring about a better understanding of the many interrelated problems and interests of the civilian and military populations and to point up the inseparable nature of the civilian-military coalition in the National Defense and security effort."[39] The seminar was attended by over 1,500 people and was held at Magoffin Auditorium at the Texas Western College (now UTEP).[40]

The Cold War therefore shaped discussion of the creation of freeways. Historian Eric Avila, who has written on urban highway construction and racial communities, states, "There was a very strong military imperative to build highways and to build them quickly because our cities are so concentrated, in the event of a nuclear attack, how are we going to evacuate our citizens from these densely, crowded, congested cities? Highways were seen as primary solution[s]."[41] The Cold War ushered in a need to protect the nation. Freeways became symbolic of infrastructure needed to mobilize the country. Few people would have taken issue with the reasons for building the interstate.

State Issues on Highway Building

John D. Huddleston in his August 1981 dissertation "Good Roads for Texas: A History of the Texas Highway Department, 1917–1947," recounts the haphazard activities of early state road building, in that before 1917 the state did not have a state highway department, and road construction and maintenance resided at the county level of government.[42] Furthermore, it was evident that state agencies did not have the resources to easily respond to the Federal Highway Act of 1956.

President of the American Association of State Highway Officials William A. Bugge, in a speech given before the Mississippi Valley Conference of State Highway Departments in Chicago, Illinois, on March 8, 1957, stated there was an urgency for state highway departments to restructure themselves to rise to the occasion like they had never done before. Bugge stated that state highway departments had failed to carry out their assignments and that they needed to carry out an integrated, long-range highway program. He told the assembly, "We can't send out a boy to do a man's job."[43] According to Bugge, two items for highway building were in short supply—skilled engineers and state money. However, Bugge's statement was indicative of the bravado of the period, and the optimism that the country would pull on its bootstraps to get the job done.

The Texas Highway Commission moved into high gear with funding for the acquisition of land from private citizens and businesses (called right-of-way) being of paramount importance. On June 20, 1956, D. C. Greer, the state highway engineer for the Texas Highway Commission, sent a letter to division heads, district engineers, and engineer managers on the creation of the right-of-way office to build the Interstate Highway System via Minute Order No. 40217. In his message he announced the creation of the right-of-way division and the selection of the division chair and outlined the department's primary functions. The State Highway Commission engaged in purchasing property to build Interstate 10, and by June 15, 1958, 2,049 parcels had been purchased for the Interstate System at the cost of $11,747,314.[44] By 1958 there were 450 full-time employees in the right-of-way acquisition section. Historically, Austin, Texas, has always had more employees than any other Texas Highway Department District Office.[45] One of the other principal issues at the state level was the proposed locations of Interstate 10 through Texas.

Local Issues: Forcing Interstate 10 to Run Through El Paso's Sand Hills

Four years earlier, on February 5, 1954, a letter written by Hosmer W. Hill to D. C. Greer, the state highway engineer of the Texas Highway Department, addressed the issue of where the freeway should be built through El Paso. Hill was opposed to the highway being built on the Bankhead Freeway or Route 80, which ran along Alameda Street. In his letter he stated, "As a private citizen, and a tax payer in El Paso County, I wish to lodge my voice against the idea of using the present hi-way [*sic*] 80 for a freeway into El Paso, regardless of what action has been taken by our Chamber of Commerce, or County Officials."[46] Hill felt conflicted about whether it was a good idea to run the new (I-10) freeway through the sand hills, which were a remnant of the continental shift, or the Rio Grande Rift, created after oceanic activity millions of years ago.[47] Hill considered the Bankhead Freeway as twenty-five years behind the times. In his letter he also stated he owned land along the sand hills route, but that that was not the reason he was writing his letter. He felt business owners along Highway 80 had long profited from the highway, while others who did not own property had not complained. To him, the route along the sand hills made more sense because he felt that "the population is moving out of the valley towards the hills, and hi-way [*sic*] 80 as it stands now would serve as a relief for the valley traffic, while the freeway in the sand hills would serve for through traffic."[48]

The route of the proposed location was a hot button in 1954. A small note was attached to Hill's letter to Engineer Greer where he explained his position. He asked Engineer Greer to reconsider and wrote, "The valley route is too crowded already, and the freeway in [the] sandhills is exactly where it belongs. Hope you will reconsider it.—H."[49]

Hill so believed in the state running Interstate 10 through the sand hills that when the time came, he donated (exchanged, according to the deed) his property for the endeavor, therefore enticing highway builders to build the highway on his land. On June 14, 1957, Hill and his wife Anna G. Hill exchanged their land in the sand hills with the El Paso County for $1 so the Texas Highway Department could build Interstate 10 through it.[50] Hill was born in 1892, and coincidentally, he and his wife lived at 3123 Durazno Street (between San Marcial and Estrella Streets), a few blocks west from Lincoln School.[51] He died February 24, 1971.

Hill's donation of this land could have been construed as a gesture of goodwill, but it might also have been a means to force the hand of the Highway Department to run I-10 through the sand hills and transfer the wealth that had developed along Highway 80 to the new area. Before I-10 was built, Highway 80 was the home to many of El Paso's most prominent families, estate homes, businesses, and hotels, all of which profited from the route. Changing the location of the freeway would drastically shift the city's wealth, spur the creation of new subdivisions and business corridors, and cause both "white flight" and the decline of stores, restaurants, hotels, and home values along the Highway 80 / Alameda Corridor.

The El Paso Highway Council

The 1955 delegation of El Paso Chamber members and officials led by El Paso Mayor Tom Rogers presented Austin officials with two route plans for the proposed El Paso freeway.[52] A *Herald-Post* article reported Mayor Rogers would not expect citizens to vote on a bond project for phase one of the freeway until the state informed citizens how much the total project would cost. The state would pay 90 percent of the cost to build the freeway, and 10 percent would be paid by the citizens via a bond vote.[53]

In 1955 the Bassett Estate owned the major portion of the sand hills freeway section right-of-way from Hawkins to the Fort Bliss spur line. City Planning Director Lowdon Wingo said there would be little difficulty in acquiring this section of land for the freeway. The article stated, "A two-mile route through this land [the airport land] was being studied by the Civil Aeronautics Administration in Washington, D.C.," and the approval for the acquisition of the land was imminent. The article also detailed that El Paso County had already acquired right-of-way for "the three miles of the sand hills route beyond airport land to the Ysleta cutoff road."[54] The Ysleta cutoff was where present-day Zaragoza Road is located today (Hosmer and Anna Hill's land).

Officials sought community input, although as evidenced by news articles, in a limited way. As a requirement of the Federal-Aid Highway Act of 1956, a public hearing was held on three links of the proposed $50 million Interstate Highway across El Paso County on Thursday, June 27, 1957, at the county courthouse. A June 22, 1957, article in the *El Paso Herald-Post* stated the meeting would be recorded and transcribed for use by officials.[55]

In 1961 the Highway Council (or the El Paso Chamber of Commerce Highway Coordinating Committee), which was comprised of business elites and El Paso Chamber members, tasked themselves with deciding on the highway routes in El Paso.[56] An April 17, 1961, issue of *El Paso Today* listed original members of the Highway Coordinating Committee as businessmen James E. Rogers, Joseph Irvin, and Edward Hines.[57] The Highway Council did not include city staff members like Jonathan Cunningham, the director of planning. The El Paso Chamber of Commerce's Highway Council was comprised of local elites, like similar groups in other parts of the country, who saw themselves as doers of the greater good, but who also had business interests.

In El Paso up to 1977, city issues were controlled by a centralized form of government and not city representatives as in the present day, so decisions to build highways and where to place them were largely in the hands of city elites and city boosters who controlled the city council and were members of the Chamber of Commerce Highway Council. As late as 1980, historian Oscar Martínez conducted a study that found few Mexican Americans in city council or among business elites or who had been directors of the Sun Carnival Parade in El Paso.[58] Attorney Tom Diamond concurred with Martinez's research and stated that when he arrived in El Paso in 1958, Anglos dominated the political arena.[59]

A January 12, 1965, memorandum for President Lyndon B. Johnson on the revised estimate cost of the Interstate Highway System stated the following:

> Attached is a copy of the report being sent to Congress today by the Commerce Department estimating the cost of completing the Interstate Highway System . . . it shows a $5.8 billion (about $18 per person in the US) increase in the total cost of the System over the 1961 estimate; The Federal share of this increase is $5.0 billion (about $15 per person in the US) . . . the revised total cost is $46.8 billion (about $140 per person in the US), and the Federal share is $42.0 billion (about $130 per person in the US), compared to the $37.0 billion (about $110 per person in the US) estimated in 1961.[60]

Each year highway building was not completed, the price increased dramatically. In a four-year period, the cost rose from $37.0 billion (about $110 per person in the US) to $46.8 billion (about $140 per person in the US), a 26 percent increase. States were required to pay a share of the cost, usually on a fifty-fifty basis. Businesses that benefited by road building in El Paso were contractors, suppliers, and businesspeople such as the Road Hands Inc., an El Paso group associated with the highway industry in 1964.[61]

Nationally, citizens who were impacted by decisions to build highways through their neighborhoods did not have a vote in their creation. Multicultural communities affected by the destruction of their neighborhoods were not part of the decision making in the creation of highways. A 2002 publication stated that highway builders knew neighborhoods and communities would be destroyed, but this was an acceptable cost.[62] Highway building resulted in the displacement of minorities along highway routes throughout the United States and was big business. Using blighted areas for higher and better uses included the creation of suburbs and business corridors, not for the greater good but for the higher dollar.

Chapter 6

"Modernizing" a Growing El Paso

Did El Paso need to be "modernized" or was it that funds were available with the 1956 Highway Act? Communities who would be affected were not at the table, nor were they invited to participate. One of the narrators said that her father, who was a building inspector, found out about building I-10 because he had seen a letter in his office at Bassett Tower about the acquisition of properties, so the community knew that the highway was coming.

In 1957 real estate brokers involved in the creation of Interstate 10 stated, "While the submerged route might cost State and Federal governments more for right-of-way it would decrease construction costs." There were varying ideas of how to build Interstate 10 through El Paso. On one hand, business owners thought that submerging the interstate through downtown El Paso would prevent blighting of adjoining properties. Yet on the other hand, in the Lincoln Park area, the decision to place freeway pillars through the community depreciated property values. Two years earlier downtown businesspeople had proposed creating a downtown bypass road to veer away from downtown like other Texas towns.[1]

Costs were also determining factors in highway construction. The Lower Valley portion of Interstate 10, consisting of a 12.2-mile section from the

El Paso junction and Hudspeth County lines to Fabens, was budgeted at $16.5 million.[2] Another 17-mile section from Fabens to the Ysleta cutoff through the sand hills was to be built by 1958, according to E. W. Mars, the state highway engineer.[3] The Texas Highway Commission visited El Paso for two days on May 5 and 6, 1958, to review plans for the freeway location.

Charting the Path for Interstate 10 Through El Paso

On May 6 the path for I-10 was announced. An article in the *El Paso Herald-Post* also offered insight on some of the politics behind the decision and Texas Highway Commission's visit to El Paso. The last event the Highway Commission held was a cocktail party and dinner at the International Club given by road and other contractors where the press was barred. The article also mentioned that Texas Highway Commission Chairman Marshall Formby would organize another public hearing in September to see if there was any opposition to the highway route, which was stated as "near Wyoming street on the north side of the present business district."[4]

Resistance to Relocation and Psychological Effects of Highway Building

Freeway Revolts began in San Francisco in the mid-1950s, and a few protests reached El Paso, such as the mural painted in Chihuahuita, a community east of Segundo Barrio in 1978. According to Chihuahuita resident Freddy Morales, a protest occurred against highway building in 1978 as depicted in a mural by Chihuahuita residents titled "Untitled" or "A Tribute to Joe Battle" or "Lágrimas" on the 200 block of Montestruc, formerly Seventh Avenue.[5] The mural was painted in 1978 by Fred Morales and Mara (full name unknown) and other Chihuahuita residents. The mural addressed the possibility of TxDOT running a freeway through the neighborhood, which happened decades later in 2019 as part of the Loop 375 Border Highway West Extension Project.[6] Protests and actions that affected highway planning would have to wait until 2006 with activities to save Lincoln Center.

Forced relocation scattered people and fragmented their kinships and connections as neighbors. Relocated individuals often became depressed and isolated and suffered psychological hardships. University of Texas at El Paso sociologist Elwyn R. Stoddard's *The Role of Social Factors in the Successful Adjustment of Mexican American Families to Forced Housing Relocation: A Final Report of the Chamizal Relocation Research Project, El Paso, Texas*, and the city of El Paso Planning Department's *Relocation Practices in El Paso*, are two important studies of El Paso relocation practices and their psychological effects.[7]

The Stoddard study includes materials collected over a five-year period, starting in 1964, from a study group of "1,155 families relocated, [and] a sample of 40 homeowners and 40 renters" who were selected and personally interviewed by him.[8] The other study, conducted by the city of El Paso Planning Department, reports that, "The Texas Highway Department became the first El Paso governmental agency in recent years to displace scores of residents with the acquisition of properties for the Right-of-Way of Interstate 10."[9]

An article titled "Stop the Road: Freeway Revolts in American Cities," documents efforts to counter highway building and describes the nature of bottom-up freeway revolts. Disagreements over property compensation did occur in El Paso, on an individual basis, and some cases went before the Justice Department.[10] William Wood recounted a story of a woman who said she was not going to give her house up. He recounts that she told Robin Washington, another appraiser, that "she was not going to give her house to those goddam Mexicans in Mexico that they could go to hell and that she was keeping her house and if necessary, she would 'get her guns out and fight.'"[11]

As a February 23, 1965, article in the *El Paso Herald-Post* pointed out, "This is the final price and if you do not sign you have 60 days from this day on and at the end of such time the amount offered to you will be deposited at a local bank and your property will be torn down whether you move or not."[12] Regardless, Lydia Hinojosa on June 1, 2015, recounted that her father did not feel he had been justly compensated for his family's business but had had no choice but to sell or face the business he had spent a lifetime building being condemned.

Highway building also dramatically changed the landscape and people's psychological sense of place. El Paso, as a small and sleepy place, experienced the beginnings of an urban transformation that continues to this day. One person lamented the change to the city in a column, "Ask Mrs. Carroll," in the *El Paso Herald-Post*, where she stated, "Dear Ann Carroll, I hate to see all these highways and freeways destroy the last vestige of sweet little sleepy El Paso." Later in her letter, she writes, "I can remember when Lincoln Park School was in the 'wilderness' and driving in the Lower Valley summer afternoons was a restful and cool respite with the smell of alfalfa blowing across your face and out past Ysleta to Socorro you could smell mesquite fires burning and see farms clear to the river."[13]

In the book *Root Shock: How Tearing Up City Neighborhoods Hurts America, and What We Can Do About It* by Mindy Thompson Fullilove on a person's dispossession due to urban renewal, she profiles the experience of a person named Dr. Claytor who spoke about his pain of having been relocated. In the passage Thompson Fullilove states, "It was the loss of a massive web of connections—a way of being—that had been destroyed by urban renewal; it was as if thousands of people, who seemed to be with me in sunlight, were at some deeper level of their being wandering lost in a dense fog, unable to find one another for the rest of their lives."[14]

Like the individuals in *Root Shock*, coming from insular and close-knit communities, students like the late David Prieto, whose family moved to the Ranchland area, where he attended Bel-Air High School, were put into a multiracial student environment in a time of turmoil. Prieto experienced racial strife and conflict and the student walkouts of the 1970s. Other displaced people suffered myriad losses, personal, physical, psychological, and economic ones. Others relocated to newer areas where they had to rebuild their communities from afar.

El Paso's (Old) East Side (African American El Paso)

The 1950s was a period when both African Americans and Mexican Americans faced disadvantages in education, housing, employment, and many other areas. African American and Mexican children endured inadequate and inferior schools and facilities.[15]

At the same time, the concentration of African Americans and Mexican Americans at the end of the city limits produced an area in Central El Paso called the East Side. What was typically called the East Side was comprised of several smaller subdivisions: Payne's Subdivision, Moeller's Subdivision, Garden Subdivision, and East El Paso Subdivision. The East Side was bordered by the present-day streets of Cotton to the west, Montana Street, to the north, Stevens Street to the east, and the Rio Grande or the US-Mexico border to the south (see fig. 8). One centuries-old community existed south of the East Side, and that was Cordova, or Cordova Gardens, as it was called.

Several racially mixed neighborhoods with large percentages of Mexican Americans and African Americans, including Lincoln Park, were part of El Paso's East Side. One of the families who lived in El Paso's East Side was the William "W. C." Calhoun Parish family, who lived on Yandell Street. In her book *Renegade for Peace and Justice*, California Congresswoman Barbara Lee states Mr. Parish became the first African American letter carrier in El Paso in 1947.[16] According to Alex Rosas, prior to the 1970s Alamogordo Street (present-day Yandell Street) located on the northern boundary of Concordia Cemetery in the Lincoln Park community was the color line: whites lived north of the street and African Americans and Mexican Americans lived south of it.[17] In fig. 12 El Paso's East Side is denoted by the broken line.

Sociologist David Montejano has argued that the segregation of Mexicans in Texas began to weaken in the 1940s due to urbanization and the rise of the Texas Mexican middle class: "Mexicans in the cities remained second-class citizens but whites were more cautious and respectful." But these social changes were not sufficient to overturn race restrictions. It took four events to challenge the racial order: "(1) the emergence of World War II; (2) the mechanization of labor-intensive agriculture; (3) the growth of urban-based commercial interests; and (4) pressure from below and outside—the civil rights struggle of the 1960s."[18] Nevertheless, according to Montejano, even before the war, in border cities like El Paso, Laredo, and Brownsville "ethnic relations were flexible and pragmatic" and "more a matter of class than of 'race'."[19]

Creating Lincoln Park, a Green Space

Under Mayor Pro-Tem Ruben Shaeffer, the city of El Paso requested from the state the use of the right-of-way beneath Interstate 10 overpass structures for the creation of Lincoln Park on August 8, 1973. Subject lands included "a total of 23 acres of land, more or less, out of and part of the right-of-way of the Interstate 10 and 110 interchange in El Paso County, Texas; said total acreage consisting of three (3) parts of areas of land herein designated as Areas 1, 2, and 3."[20] City Planner Nestor Valencia drew up plans for the park.

Starting in 1974 Lincoln School was considered for use by El Paso Community College (EPCC). In a May 1975 letter to Jonathan R. Cunningham, director of planning and research for the city of El Paso, Assistant District Attorney Bruce Yetter stated that the city would convert the twenty-three acres into public use and that they would renovate the old Lincoln School.[21] In an e-mail titled "4001 Durazno Timeline" to El Paso Quality of Life Director Deborah G. Hamlyn, El Paso City Engineer Alan Shubert stated, "[On] August 8, 1974, the city passed a resolution authorizing the mayor to sign the Multiple Use Right-of-Way Agreement with the State of Texas authorizing the city to use State-owned right of way beneath Interstate Highway 110 structures for public purposes."[22] Funds from the Housing and Urban Development Act of 1974 were used to upgrade the Lincoln School and reopen it as a cultural arts center.

A year later the El Paso Metropolitan Planning Organization (MPO) was created. The El Paso MPO was created on June 23, 1975, at an El Paso City Council meeting under Mayor Don Henderson with the appointment of Mrs. Judy Price to the position. An MPO is a local decision-making body that is responsible for overseeing the metropolitan transportation planning process. MPOs are required for urban areas with a population of more than fifty thousand people.[23]

In September 1975 a resolution was signed authorizing the mayor to sign a nonalienation agreement with the Economic Development Administration for remodeling the Lincoln School building and turning it into a recreation center. The building was remodeled in 1976 with $100,000 of Community Development Block Grant (CDBG) funds. The center would house offices for divisions of the Parks Department (later renamed Parks and Recreation Department, or PARD) and would provide spaces for cultural events for the Parks Department.[24]

Remodeling Lincoln School was part of Mayor Fred Hervey's 1975 Community Development Program funded by the Housing and Community Development Act of 1974.[25] Mayor Hervey convened city-wide meetings with El Pasoans to decide on projects for the program. Over four thousand citizens attended and identified funding priorities in their neighborhoods.

The September 1975 agreement made it possible for the city to retain the title to the Lincoln School and have a leasehold interest for the useful life of project facilities, which was determined to be forty years. The agreement included waivers of, and modifications and alterations to, restrictions and limitations, which required prior written consent from the assistant secretary of commerce for Economic Development in accordance with Title 13, Code of Federal Regulations. The contract expired in 2015, and the center was returned to TxDOT that year.

The I-10, US 54 Interchange

The Lincoln Park community acted as ground zero where all four of El Paso's highways would meet in the city's center. El Paso's highways (those built between 1960 and 1970) converged at an interchange known as the I-10, US 54 Interchange, which was deemed a state-of-the-art structure when it was built in 1970, and which became known as the Spaghetti Bowl. The "ultra-modern, four-level interchange was built between December 1968 and April 1969 for $17,000,000."[26]

At the time of its construction, the I-10, US 54 Interchange was the largest highway interchange project ever undertaken by the state of Texas, and the most complicated to build. The design consisted of 858 pages of blueprints. Lyle D. Scarbrough, a supervising design engineer for District 24, stated, "The last sheet should have been specifications for a cart to carry them around." Material needs to build the structure required 48,000 cubic yards of excavation, 62,000 cubic yards of concrete, 12.45 million pounds (about 5,647,220.4 kg) of reinforcing steel, 6.4 million pounds (about 2,902,988.8 kg) of structural steel, and more than 22 miles of prestressed concrete beams.[27] As a physical structure, the Spaghetti Bowl represented a technological marvel and signaled other developments in highway building to come.

The Bridge of the Americas and the Programa Nacional Fronterizo

The push for urban renewal was not the only factor influencing the creation of Interstate 10 and the Spaghetti Bowl. Highway building made it possible to connect with the proposed BOTA, which was part of the Chamizal Settlement with Mexico. According to Mexican urbanists Marisol Rodríguez and Hector Rivero, the Mexican government initiated the National Border Program, or ProNaF (Programa Nacional Fronterizo), hoping to align itself with the urban changes taking place on the US side of the border in the 1960s. One description of ProNaF was "a regional and urban planning initiative to face an economic openness in the southern and northern border cities which aimed to detonate an integrated development model."[28] Art historian Amy Sara Carroll offers another description of the project by stating it was "barbaric, empty, seditious, illicit, and Americanized, in a DF-centric imaginary."[29]

Two Mexican architects, Mario Pani (1911–1993) and Domingo García Ramos (1911–1978), working in connection with the architectural school El Taller del Urbanismo, developed urban plans for the project. In addition, the ProNaF project answered the call for an "urban border proposal" for the land that was returned to Mexico via the Chamizal Settlement. Not wanting to just build highways, developers of ProNaF worked with the Technical Council of Regional Planning and proposed building a superblock of urban cells with freeway access on the Chamizal property (a very different approach to freeway building than that adopted in the United States).[30] The project was an effort to modernize the city of Ciudad Juárez and change the notions of what the city and the border meant to Americans.[31] Carroll states that before ProNaF, "The prevailing image of the border was one of anarchy—illicit pleasure, transience, lawlessness smugglers' and informal economies—matched by the free spirit and barbarism of the Wild West."[32] Wider streets, designed to attract American tourists and make them feel like they were in the United States, were incorporated into the proposal.[33]

During this period Mexico sought to create linkages to US cities all along the border for trade purposes, as well as for tourism and industrialization. ProNaF was used as a test model for a commission on border urban development known as the Mixed Committee on Border Urban Development (Comisión Mixta de Desarrollo Urbano Fronterizo), which via regulatory plans (*planes reguladores*) created master "urban development plans for several cities in the northern Mexican border including Ensenada, Tijuana, Nogales, Mexicali, Piedras Negras, Matamoros and Ciudad Juárez.[34] These regulatory plans laid the foundation for the modern maquiladora industry. Maquiladoras began operating in Mexican border cities in 1965.[35]

These mid-1960s modernization efforts may have been successful at the onset, but over time they became obsolete. Sixty years later a project to widen the BOTA was announced as an effort to modernize and humanize southern ports of entry and utilize funds from both sides of the US-Mexico border with the participation of President Biden's infrastructure bill with support from the governors from the states of Chihuahua and New Mexico.[36]

Demolition of El Calvario Catholic Church

In 1968, while Mexico was creating a utopia for American shoppers and preparing itself for a future of trade with the United States, Interstate 10 and Highway 110 were rising around El Calvario Catholic Church and Lincoln School. The footprint of El Calvario, a stone church built in 1933, located at the corner of Durazno and Martínez Streets, was needed to make way for a freeway column to support a section of Highway 54 leading to the BOTA Port of Entry. In 1970 El Calvario, the church where El Paso County Judge Richard Samaniego had served as an altar server and that was the pride of the Lincoln Park community and Catholics in surrounding neighborhoods, was demolished. What is problematic about the destruction of El Calvario was that there had been no governmental study as mandated by present-day agencies such as the National Historic Preservation Act, Section 106 Process.

The National Historic Preservation Act was created in 1966. The Section 106 (16 USC. 470f) process states, "The head of any Federal agency having direct or indirect jurisdiction over a proposed Federal or federally assisted undertaking in any State and the head of any Federal department or independent agency having authority to license any undertaking shall, prior to the approval of the expenditure of any Federal funds on the undertaking or prior to the issuance of any license, as the case may be, take into account the effect of the undertaking on any district, site, building, structure, or object that is included in or eligible for inclusion in the National Register." Further, it states, "The head of any such Federal agency shall afford the Advisory Council on Historic Preservation established under Title II of this Act a reasonable opportunity to comment with regard to such undertaking."[37]

There was no historic resources study of El Calvario in 1970. Forty-eight years later, in 2018, it would be different scenario due to the activism and resistance of the Mexican American community concerning the possible demolition of Lincoln Center due to the I-10 Connect Project.[38] And then in 2024, for the first time, El Paso's Black community became involved in the Section 106 Process for the Reimagine I-10 Project, having been apprised of the need to protect and advocate for the areas in proximity to their churches and the remnants of their once-thriving business and walkable community.

At the church service on that fateful day, Father Pacheco began the mass in a solemn and emotional state and declared it was going to be the last mass to be held, and that the state needed the church, and it would be torn down. Witnesses state Father Pacheco's voice broke, and his eyes welled up with tears. Parishioner and Lincoln Park resident Teresa Orozco was there when Father Pacheco made his announcement. She said he began giving the church saints and objects away to parishioners at the mass. She recalled many parishioners crying during the service. The choir sang, and the last mass was held in Latin. Before the church was torn down, the altar was moved to Guardian Angel Church at 3021 Frutas, and church statues were given to devoted families. The life-sized Catholic crucifix with the tableau was moved to Guardian Angel Church at 3021 Frutas Street, where it currently resides. The tragedy regarding the demolition of El Calvario Church was the National Historic Preservation Act was in place as early as 1966; but in 1970 when the church was demolished, parishioners and Lincoln Park residents were not notified by the Highway Department.[39]

Lincoln School itself was also closed in 1969, and the city of El Paso sold the school to TxDOT for $20,000. East of the Lincoln Park community, suburbs and shopping malls were being built as early as 1961 (such as the Bassett Place shopping mall) while the roads were being cleared for the creation of Interstate 10.[40] During this period, Montana Street was also widened. Part of Stephenson Street in Lincoln Park was removed in 1973, when the city of El Paso requested access to the space beneath 1-10 to develop open space for public use, so present-day Lincoln Park could be created. The Highway Department purchased properties via eminent domain for the construction of the ramps and lanes connecting Highway 54 with the BOTA and Interstate 10.

The city of El Paso was granted access to the state-owned right-of-way beneath the I-10 overpass structures for public use. Joe Battle, in a letter to Mayor Paul Hervey, stated, "This office has examined the agreement between the City of El Paso and the State of Texas concerning the multiple use of land in the IH-10 and US 54 Interchange and has determined it is possible to obtain approval of the City's request for the El Paso Community College to use the Old Lincoln School building, provided an agreement executed between the City of El Paso and the El Paso Community College is acceptable to the State and Federal Highway Administration."[41]

In 1974 even though El Calvario Church was closed, residents continued having church services in the park under the canopy of the freeway overpasses. Orozco said former El Paso County Commissioner Orlando Fonseca helped open Lincoln Cultural Center so the residents could hold services at Lincoln Park after the church was demolished.

Purchasing Property for El Paso's Highways

In addition to Interstate 10, other freeways around the Lincoln Park community further reshaped its space. In 1969 EPISD sold Lincoln School and its adjoining twenty-three acres to the Texas Highway Commission for use as a field office for overseeing the construction of Interstate 10, Gateway East and

West, and the Spaghetti Bowl, and as a staging location for construction and maintenance equipment and personnel.[42] In December of 1968, Battle, the District 24 engineer, said Interstate 10 was mandated to be completed within a period of two years.[43]

The property to build I-10 was acquired by the county of El Paso as early as 1954, two years before the passage of the Highway Act. The purchasing of properties for highways continued in the state for some time. In January of 1964, the Texas Highway Commission voted on the acquisition of parcels for the construction of Interstate 10.[44] Between January 29, 1964, and August 31, 1967, the commission purchased forty-three parcels to build Interstate 10. Between August 31, 1967, and January 1970, the commission purchased an additional thirty-eight parcels to build Highway 110, and on January 29, 1970, it purchased two parcels to build Loop 375. Parcels for the creation of US 54, the North-South Freeway, and parcels to build the Border Freeway were purchased by the federal government. The creation of the Chamizal Freeway was coordinated by the Texas Highway Department of the US Department of Transportation project.[45]

Table 6.1 State Highway Department Property Purchases Between 1950 and 1970 by Subdivision

Addition	Properties Purchased
Azcarate	1
Bassett Addition	1
Beaumont	82
Brentwood Heights	43
Buena Vista	21
Campbell	48
Chula Vista	2
Cotton	3
Cordova Gardens	2*
Dichiara	40
E. Bennett Survey 12	2
E. Dennett Survey 11	2
East El Paso	3

(Continues)

Table 6.1 Continued

Addition	Properties Purchased
El Valle	5
Franklin Heights	21
French Addition	111
Globe	1
Government Hill	185
Grandview Addition	17
HA Chadwick Sur 253	1
John Barker Survey 10	4
Latta's Woodlawn	52
Lincoln Park	23
Logan Heights	51
Magoffin	5
Mesa Heights	38
Mobile Place	1
Morehead Block 1	2
Morningside Heights	14
Mundy Heights	23
Old Fort Bliss	5
Orndorff	1
Park 2	2
Paynes	31
Rivera	32
San Elizario Grant	4
Satterthwaite	13
See Instrument	22
Skyline Hills	23
Socorro Grant	1
Stevens	2
Sunrise Acres 1	26
Sunrise Acres 3	16
Sunset Heights	30
Washington Park	19

(Continues)

Table 6.1 Continued

Addition	Properties Purchased
Westlyn Heights	12
Wilkins	6
Womble Addition	12
Zia Village	3
TOTAL	1,064

Note: * This number does not include the fifty properties in the Rio Linda Subdivision that were returned to Mexico because of the Chamizal Settlement.
Source: County of El Paso, Texas, Historic Deed Index Books, https://kofilequicklinks.com/ElPaso/Default.aspx. The online index was created under the tenure of Delia Briones, El Paso County Clerk.

An analysis of deed records from the El Paso County Records identifies the purchase of properties by the State Highway Department between 1950 and 1970 by subdivision: As the above chart illustrates, highway building affected numerous subdivisions but most significantly the Beaumont Addition (82 properties purchased); the Campbell Addition (48 properties purchased); the French Addition (111 properties purchased); the Government Hill Addition (185 properties purchased); Latta's Woodlawn (with 52 properties purchased); and Logan Heights (with 51 properties purchased). In contrast, only 23 properties were purchased in Lincoln Park. Brentwood Heights had 43 properties purchased; and Dichiara had 40 properties purchased. The other subdivisions on the list lost fewer than 40 properties each. A total of 1,064 properties were purchased.

On average, families in the Lincoln Park neighborhood were paid from $4,000 to $6,000 for their homes. Prices were decided by the lot size or number of residents owned, not by the physical characteristics of homes on the lots. Where families relocated to depended on how much money they received for their homes. Many families moved into communities where they could purchase homes for the amounts they were paid.

As late as 1968, a mortgage for a three-bedroom, two-bathroom cinder-block home with a front yard and a large backyard in the Hacienda Heights area ran for $10,000, with a payment of $100 a month. Many families moved into homes built by Hervey Construction (later the Hervey Real Estate Company), which had converted the land south of the sand hills into the new

East Side neighborhoods, where the future Interstate 10 and its gateways would eventually run through.

However, like other communities of color in the country and in Texas, most of the El Paso communities affected by the creation of Interstate 10 and other freeways were not represented in the decision making. City politics in the years from 1954 to 1968 (the years Interstate 10 was built) was controlled by elites who did not represent El Paso's Mexican American and African American citizenry. Further, since road building was linked to military efforts, few citizens fought against the building of the freeways.[46] Despite the devaluation of properties and the characterization of creditworthiness of African Americans and Mexican Americans during this period, community churches acted as important anchors in the lives of residents. A common thread among the oral histories of residents who lived in the Lincoln Park Community was the key role of their neighborhood social groups in keeping their community bonded.

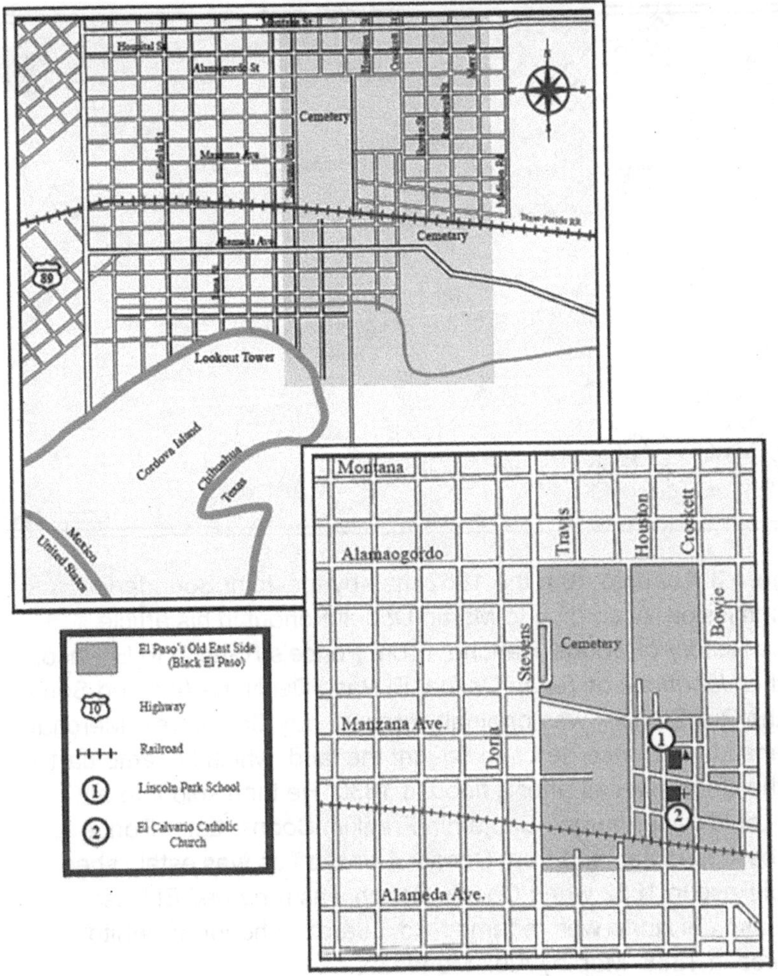

Figure 2 The Village of Concordia, Camp Concordia, and Concordia Cemetery. The present-day boundaries of Concordia were Montana Street to the north, the Rio Grande surrounding Cordova Island to the south, Stevens Avenue to the west, and Marr Street to the east. Camp Concordia was located inside Alamogordo to the north, Frutas to the south, Stevens to the west and Bowie to the east. Map by Alex Mendoza.

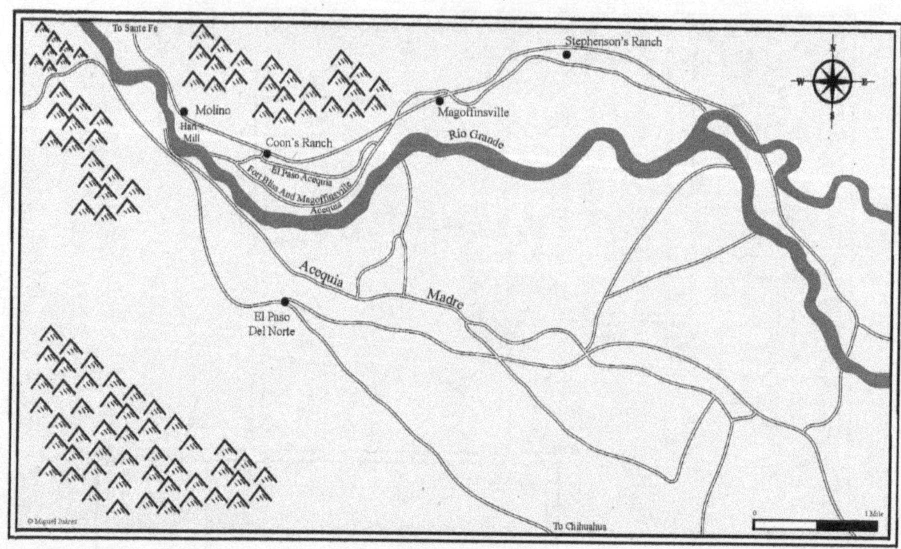

Figure 3 Adapted from the 1852 map by the Joint Boundary Commission. According to Martin Donell Kohout in his article "The History of Coons' Rancho: From Ponce's Rancho to El Paso," in the *Handbook of Texas Online*, El Paso Del Norte (located South of the Rio Grande) was originally present-day Cd. Juárez. Merchant Juana Maria Ponce de Leon bought the land, which became part of the United States after a flood in 1830. He then sold it to a St. Louis trader named Benjamin Franklin Coons. Later Ponce repossessed the land from Coons. A post office was established in El Paso in 1852 when Coon's Ranch was renamed El Paso. El Paso del Norte was renamed Cd. Juárez in honor of Benito Juárez in 1888. Map by Alex Mendoza.

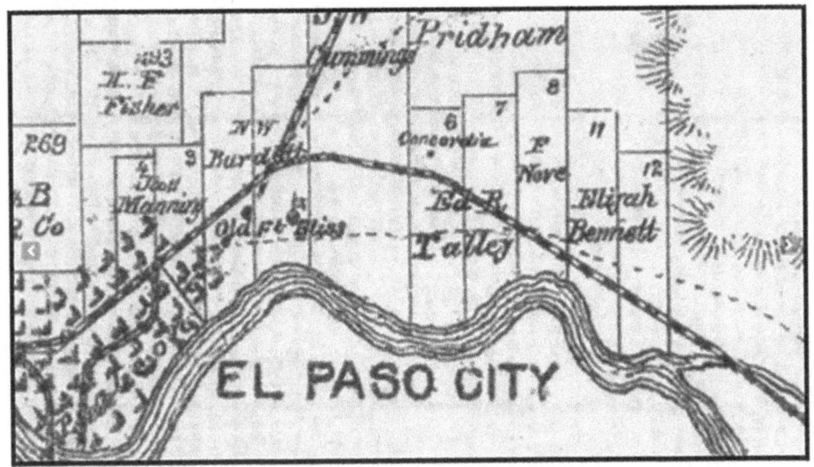

Figure 4 Detail of Map of El Paso County, Texas, 1893-94. C. J. Canda, S. J. Drake & W. Strauss, Texas & Pacific Railway Company, W. H. Abrams, General Agent, Dallas, Texas. Contributor: M. Stakemann, Texas & Pacific Railway. Library of Congress.

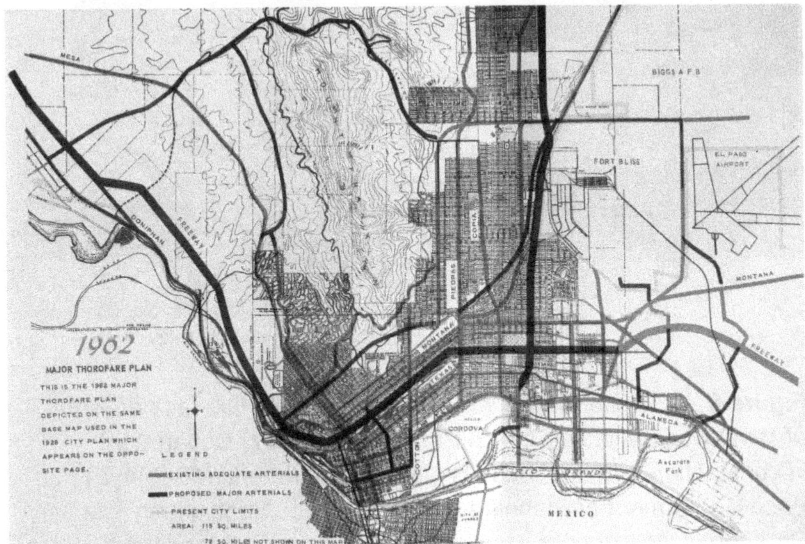

Figure 5 The Major Thorofare Plan, 1962. This map used the 1925 Kessler Plan as a base map. It placed secondary arterials (highways) and major arterials throughout the city. MS 204 El Paso Department of Planning Collection, Box #59, Folder 32, p. 51, C. L. Sonnichsen Special Collections Department, University of Texas at El Paso Library.

Figure 6 *La Compra Del Derecho De Via* (*The Purchase of Right-of-Way*) was printed by the El Paso City Planning Department in conjunction with the Texas Highway Department. MS 204, El Paso Department of Planning Collection, C. L. Sonnichsen Special Collections Department, University of Texas at El Paso Library.

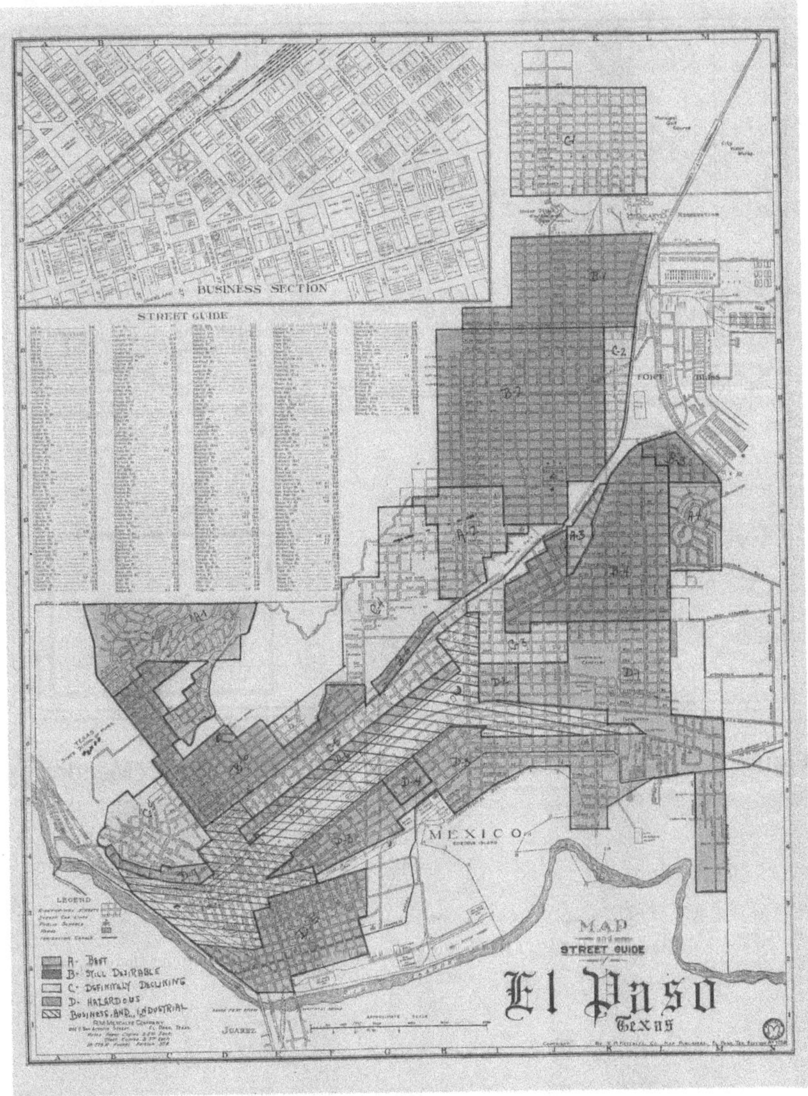

Figure 7 El Paso "Residential Security" map. Record Group 195: Records of the Federal Home Loan Bank Board, 1933–1989, Box 154, folder for El Paso, Texas, National Archives and Records Administration, College Park, Maryland.

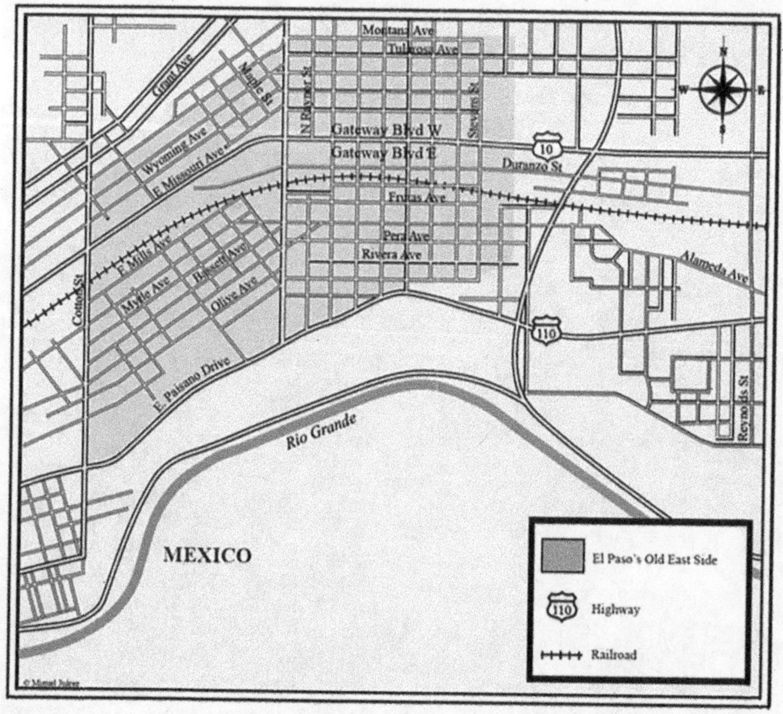

Figure 8 El Paso's East Side (Black El Paso). Map by Alex Mendoza.

Figure 9 *Virgen de Guadalupe with Roses*, mural by the late Felipe Adame, 1981. First mural painted at Lincoln Center. Photograph by the author.

Figure 10 Texas Highway Department District Office 29 (El Paso) office members, 1969, standing in front of maquette of "Four-Level Highway Interchange" (locally known as the Spaghetti Bowl) and an almost two-foot stack of plans (held up by three benches). Manuel Aguilera is standing in the front row (*left side*) wearing a dark turtleneck shirt and moustache. Photograph courtesy of and with permission from the late Manuel Aguilera.

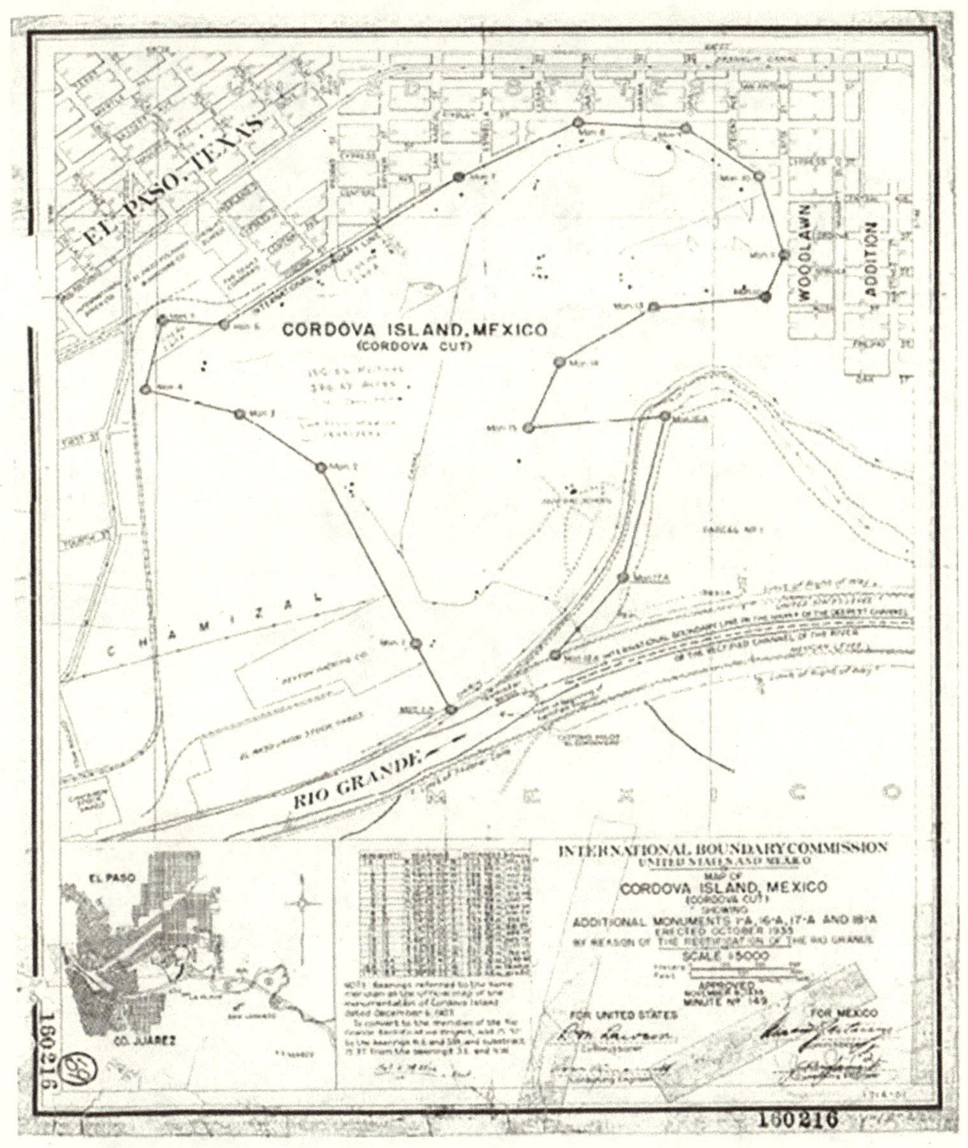

Figure 11 Plat Map of Cordova Island, Mexico. City of El Paso Planning Department.

Figures 12 and 13 Community members protesting demolition of Lincoln Center. Photograph with permission of Federico Villalba, whose family was displaced by the North–South Freeway.

PART 3

Resistance and Renewal

Chapter 7

Speaking Truth to Power

Lincoln Center, Lincoln Park, El Corazón de El Paso (1970s–Present)

This chapter details the process of how a lowrider car club became a neighborhood association and later a preservation organization that advocates for the community and the space known as Lincoln Park in South Central El Paso. The chapter is divided into two sections: the first relates developments that led to the closure of Lincoln Center and the formation of the LPCC, an organization that advocated for the reclamation of communal space and organized community actions and events; the second tracks specific actions the LPCC took to engage with the community and prevent Lincoln Center from being demolished. It examines the David and Goliath effort on how a grassroots Chicano/a organization with the support of the community and key politicians and media support and media opposition rose to meet the challenge of keeping Lincoln Center from being demolished. It also argues that Lincoln Center and the Lincoln Park community, which was the site of the third Fort Bliss, has been overlooked as a historical, cultural site.

It is important to note that the neighborhoods surrounding Lincoln Park—such as San Juan, El Sovaco, La Roca, Washington Park, and Val Verde—have not been studied historically as much as other more affluent parts of the city. It is imperative that these communities are studied because they are in the process of being gentrified, not block by block, but housing lot by housing lot, to developers who insist that they are building housing for the emerging hospital complex. In a 2018 article in the *El Paso Times* reporter Vic Kolenc reported that the Medical Center of the Americas had identified 140 acres (about twice the area of a large shopping mall) for its medical center in South Central El Paso.[1]

There are a multitude of reasons why lower income communities' histories are overlooked. It could be that long-time residents have moved or passed away, or that their children have moved away, or that politicians who were sympathetic to residents have passed away and stopped advocating for residents. Also, historic preservation in Latino/a communities is a more recent development. In 2011 the LPCC learned that the old Lincoln Park School building was to be demolished. In 2014 a series of events galvanized the committee, the community, and local politicians. From then on Lincoln Center was seriously taken into consideration as a cultural heritage site that needed to be preserved and, if possible, reopened.

The Lincoln Park Cultural Arts Center (Lincoln Center)

A report by the city of El Paso states that CDBG stated funds were to be used to renovate Lincoln Center: $100,000 in 1976 to renovate the center; $56,898 in 1978 to install an elevator; $24,950 in 1987 to replace the roof; and $29,000 in 1988 to repair and renovate bathrooms.[2]

In 1980 the city opened Lincoln Center to arts groups. Lincoln Center had offices for Parks and Recreation Aquatics Division, Sports Division, and after-school services. It also hosted various nonprofits including LULAC Project Amistad and Project Bravo. Organizations such as the Juntos Art Association and others sponsored art exhibits.[3] Artists continued having exhibited in the gallery. The Lincoln Art Gallery hosted about three thousand city artists and children from 1981 to 2006. It operated a daycare and had an auditorium for performances and gatherings.

Bobby Adauto, then the director of the Lincoln Center, asked artist Felipe Adame to paint the first murals on the pillars under the Spaghetti Bowl by

Lincoln Center in 1981. In 1981 he painted a mural featuring the Virgen de Guadalupe. Later, in 1983, Adame added roses to the base of the mural and renamed it *Virgen de Guadalupe with Roses* in remembrance of his son (see fig. 9). Coincidentally, Adame's mural was on the pillar where El Calvario Church once stood.

In addition to the exterior murals at Lincoln Park, there are several important murals painted inside Lincoln Center. One is titled *Tribute to Abraham Lincoln*, painted by Artist Carlos E. Florés in 1984. Another is titled *Amistad/ Friendship*, also painted by Florés, in 1985, and assisted by Fermin Montes, Manuel Guzman, Ana Ramos, Flaviano Ortíz, Fernando Galvan, Carlos Casillas, and Enrique Florés. They were funded by the City Park and Recreation Department and the Upper Rio Grande Private Industry Council (PIC). Florés attended La Academia de San Carlos in Mexico City, the first major art academy and the first art museum in the Americas. He studied painting under Maestro Luis Nishizawa. Nishizawa is recognized as one of Mexico's leading landscape artists of the twentieth century. La Academia de San Carlos is the same university attended by Mexican artists like Saturnino Herrán, Roberto Montenegro, Diego Rivera, and José Clemente Orozco.

The late Lucy Acosta, a noted community leader and a member of LULAC, opened her office, Project Amistad at Lincoln Center in 1976, the day the Lincoln Cultural Center was reopened after its remodeling. The city of El Paso rented her space for $1 a year. Acosta was the first woman appointed to the Civil Service Commission of El Paso and was inducted into the Texas Women's Hall of Fame in 1987. Acosta's life is chronicled in "Abuelitas de El Paso: A Selective Cultural Story," 1989 master's theses by Lilia Patricia Reade-Pellicano as part the master of arts in Interdisciplinary Studies at UTEP.[4] LULAC / Project Amistad was a program that worked and advocated for the rights of elderly and disabled populations. Humanitarian Rosa Guerrero housed her Ballet Folklorico dance company at Lincoln Center on two separate occasions, first in 1975 and later in the early 1980s.

The Lincoln Art Gallery presented the *Juntos 1985* First Invitational Hispanic Art Exhibit, September 13–October 11, 1985, organized by Paul H. Ramírez and the author. The exhibit featured prominent artists from El Paso and Cd. Juárez, including the late Manuel G. Acosta, the late Rudy Montoya, the late Luis Jiménez Jr. (who created the *Los Lagartos / Alligators* sculpture in San Jacinto Plaza), the late Marta Amaya Arat, Ernesto P. Martínez, Mago Orona, Antonio Piña (who had been a student at Lincoln Park School), Paul H. Ramírez, the author, and Cd. Juárez artists: Noel Espinoza, Miguel Varela, Alicia Acosta De Sanz, Ildefonso Bravo, Velia Carranza, Elvira Fe de Mirano, Rebecca Antuna, Antoñio Arrellanes, Irma Camacho, and Lucina Chavéz.

The Juntos exhibit led to the creation of the National Association for Chicano Arts (NACA) that later changed its name to the Juntos Art Association in late 1986. NACA was conceptualized by artist and designer Antonio Piña, who had been invited to exhibit as part of the National Association of Chicano and Chicano Studies (NACCS) Conference, which took place in El Paso, Texas from April 10 to 12, 1986.

After participating in the NACCS Conference in El Paso, Piña felt a national association of Chicano Arts was needed. Many of the artists who exhibited in *Juntos 1985* formed NACA and later the Juntos Art Association, currently in its thirtieth year of existence. In late 1985 there were also meetings at the homes of Ernesto P. Martinez and Piña to solidify the group that was the first Mexican American Artists' Group in El Paso.

El Paso artists and photographers (including David Nakabayashi and Gabriel Gaytán) had their first shows at the gallery, and numerous performance groups performed in the amphitheater next to Lincoln School. In 1986 the gallery was the site of the *Juntos 1986: Hispanic Photographers* exhibit that included notable Latino photographers, including internationally recognized photographer Carlos Fernández. It is estimated that from 1981 to 2006, the Lincoln Art Gallery hosted about three thousand local children. Lincoln Center served the Lincoln Park neighborhood and residents of the Durazno and Chamizal neighborhoods.

The Closure of Lincoln Center

The turn of the twentieth century saw the decline of population in the Lincoln Park community. A major flood in 2006 was blamed for the closure of Lincoln Center. The LPCC emerged, took ownership of the open space next to Lincoln Center, beautified it, and converted it into El Paso's Chicano Park. In an interesting turn of events, LPCC began organizing events at Lincoln Park, which had long thrived as a community gathering place following its creation but by 2006 had been forgotten.

LPCC members found a model to refer to in considering potential uses for the space when Hector González, his brother David González, and their family visited Chicano Park in San Diego in 2008.[5] When they returned to El Paso, the González brothers sought to emulate what they had seen in San Diego's Chicano Park, which led to the establishment of Lincoln Park Day. The event grew every year, and LPCC members became motivated to research the rich history of the community that had been lost due to highway construction and neglect by the city. Each year bits and pieces of Lincoln Park's history were rediscovered until its importance could no longer be disregarded by the city and state officials, and the community's quest to reopen the center could no longer be ignored. This story gives a face to the community's unknown history.

From 2005 to 2007, the Latin Pride Car Club organized an event called Arte en El Parque / Art in the Park, modeled after the city of El Paso's Art in the Park but with a Chicano twist.[6] In 2008 the event became the annual Lincoln Park Day, modeled after similar events held at Chicano Park in San Diego. A year earlier, in 2004, Hector González, firefighter and president of the Latin Pride Car Club, had e-mailed the city of El Paso about reopening Lincoln Center. His email went unanswered.

Heavy Rains and Summer Floods in 2006

In the summer of 2006, heavy rains descended on El Paso and two months later, in August 2006, torrential rainfall caused flooding in El Paso and the Lower Valley. South Central El Paso, the Lower Valley, and rural areas suffered the worst of it. Funding for flood recovery in the city historically has been uneven, and as anthropologist Timothy W. Collins states, "Within the City of El Paso, investment of the social surplus by the local state in flood recovery has been uneven, having facilitated elite geographical groups of people on the Westside at the expense of less powerful socio-spatial groups."[7] According to El Paso City Manager Joyce Wilson, Lincoln Center was "severely impacted by water damage."[8] Following the 2006 floods, the city shut the center down with plans to reopen it.

Coincidentally, González's fire unit had been called to assist Saipan-Ledo residents in evacuating their homes during the floods.[9] Lincoln Center was used as a rescue station for flood evacuees from the Saipan-Ledo neighborhood, where fifty-six homes were destroyed. A recurring theme in this book is the marginalization of households in lower economic areas by land clearance for road building or natural disasters such as flooding. It is important to note that the damage to homes in the Saipan-Ledo neighborhood, resulting from the breach of the pond, did not affect Lincoln Center. Lincoln Center did not get flooded as city officials claimed but instead, as related above, was used as an evacuation center. In October 2006, citing mold issues caused by flooding, city officials closed the center. There were mold issues, but they had been caused not by flooding but by the city's negligence and improper installation of air conditioning units. Thereafter, the city inspected the building to determine if it could be reopened or demolished. LPCC members felt the inspection was intended to justify the building's demolition.

The Microbial Investigation on Lincoln Center was conducted by Sun City Analytical Inc. on October 16, 2006. Based on their report, the city's Environmental Services determined that the center should be closed temporarily for the mold to be cleaned out and structural deficiencies repaired. The city expressed plans to reopen Lincoln Center in November 2006, but it never did. The *Microbial Investigation* report, upon close examination, reveals the main reason for water intrusion into the Lincoln Center to be not flooding but "window caulking that was missing or damaged, roof and parapet wall damage, and/or exterior plaster cracks."[10]

The inspection report states that the main cause of "moisture in the building [was] the evaporation cooling system" which was "building excessive moisture in the air plumes due to missing or inoperative exhaust systems." The evaporation coolers were only two years old. The report stated that they should not have been used in a building exposed to high volumes of soot and tire dust from the freeway. The report stated that no mold was found in the ducts, but that soot and rust were found on the evaporation cooler. The report recommended that refrigerator coolers should have been used instead. Testimonies by residents who experienced the floods firsthand concurred with the *Microbial Investigation* report that there had not been damage to the carpet in the building, and flooding had not reached Lincoln Center.[11] The majority of the mold growth was found on ceiling tiles. The report did not recommend demolishing the building.

According to City Manager Joyce Wilson and the *Microbial Investigation* report, the closure of the building was ordered by Dr. Jorge Magaña, director of the City-County Health District, "who requested the closure of the Center and that all personnel vacate the building due to (1) that asbestos . . . with positive aspects of mold were present in the building; (2) that the investigation revealed air contamination with mold permeating throughout the entire building; and (3) that Center staff members had reported respiratory ailments due to contaminants."[12] The building has remained closed since 2006.

Attorney Ray Eli Rojas submitted several Freedom of Information Act (FOIA) requests to the city. The documents he obtained confirmed the city's intent to tear down the center. The documents revealed that city officials had been planning the demolition since 2007 and had not notified residents of their intentions.[13] In an e-mail dated January 2, 2008, from former City Manager Wilson to Deborah Hamlyn and Nanette L. Smejkal to her staff, she states that Wayne Thorton had made some public comments about Lincoln Center and the flooding and relocation, and she wanted to reinforce the idea that they had reached consensus in the reopening of the facility—that it would not be reopened and that instead they had plans to use the Y (what later became the Pat O'Rourke YMCA) building purchase to take the place of Lincoln Center.[14] Wilson's e-mail to her staff stated, "Make sure we are speaking with one voice and that he understands that his operation is not planned to go back there, Thx."[15]

Wayne Thornton, who had long managed the city's Youth Recreation Program, had inadvertently made comments to local newspapers about his office returning to Lincoln Center. Wilson, who had closed the center in 2006 for no other reason than wanting to save money for the city, was vehemently opposed to the idea. In her e-mail message concerning Thorton's comments, Wilson stated, "Nanette: Please find out why this happened?

THIS FACILITY SHOULD BE CLOSED—PERIOD. We all agreed on that, months ago, thanks."[16]

Wilson might have wanted to close the center to divert funds elsewhere, but the reason she gave, that Lincoln Center was flooded during "Storm 2006," had been disproven two years earlier by Sun City Analytical inspection.[17] As Collins and other authors quoted in this research illustrate, it is easier to marginalize communities that are vulnerable to socioenvironmental changes without their input in decision-making processes.[18] While deliberations regarding whether to demolish the Lincoln Center progressed inside city hall, the LPCC painted murals on the pillars at Lincoln Park and continued holding annual Lincoln Park Day celebrations. These actions amounted to the LPCC claiming the park and Lincoln Center as a place of Chicano/a culture, despite the building being closed.

Creation of the LPCC

In 2008 brothers Hector and David González and artist Gabriel Gaytán created the LPCC to advocate for the needs of Lincoln Park.[19] In 2009 a mural titled *El Corazón de El Paso*, was painted on a thirty-by-twenty-foot T-shape freeway column by artist Gabriel S. Gaytán. Gabriel's son, Gabriel Itzai, assisted Gaytán in the painting of the mural, as did the members of the Latin Pride Car Club, who also contributed scaffolding and materials. The idea for the iconic mural came from David González, a member of the Latin Pride Car Club, who commissioned and sponsored the mural. González gave a preliminary sketch of his idea for the mural to Gaytán, who added images of the Franklin Mountains, the star on the mountain, and Mexican pyramids to the composition to symbolize El Paso's over 85 percent Mexican American heritage and population.

In 2010 LPCC joined with residents to create a neighborhood association, and with the Partners-in-Parks program, which made it possible for them to sponsor four events a year: César Chávez Day in March, Lincoln Park Day in September, Día de los Muertos in November, and Día de La Virgen de Guadalupe in December. Proceeds generated from these events offset the expenses of organizing the events, the park's beautification, and the painting of murals by local artists. The Lincoln Cultural Center and Park has remained a social anchor and meeting place for the community, and meetings and ceremonies are often held outside of the closed center in the amphitheater and park areas. One of the goals of LPCC was to collaborate with the city of El Paso to make Lincoln Park a cultural tourist destination for the arts. In partnership with the El Paso Convention and Visitors Bureau, in early 2010 the LPCC designed, and the bureau printed, ten thousand full-color brochures to promote awareness of the murals and the park. The brochure was designed by artist Gabriel Gaytán.

LPCC's Campaign to Save Lincoln Center from Demolition

In 2010 because of the effort to keep Lincoln Center and Lincoln Park from being demolished to make way for the I-10 Connect Project, the Lincoln Park community became the epicenter for citizen activism and resistance to highway building. Neighborhood and local communities' efforts were responsible for the eventual transformation of Lincoln Park into a site of cultural and ethnic pride. The LPCC led the effort against the destruction of Lincoln Park. As mentioned previously, the Latin Pride Car Club, which has the same membership and acronym as the Lincoln Park Conservation Committee, had sponsored the annual Lincoln Park Day since 2005. Every year since its closure in 2007, LPCC had asked itself why Lincoln Center had not reopened.[20]

In 2011 TxDOT requested the demolition of Lincoln Center by the city of El Paso because it "determined that the Center is not eligible as a historical structure nor its use contemplated in the future."[21] TxDOT did not see Lincoln Center as a historical site. The city echoed this view by sending their historic preservation officer to conduct an analysis of the center, and they concluded in an email that among other issues with the building, there was "no information on anyone of prominence that had attended the school" nor "information on anything of significance that happened in the building."[22] These statements were unfounded.

This book provides substantial proof of the historic importance of the site and adds to a growing body of work uncovering histories of other urban communities of color. These are the challenges of writing the histories of multicultural urban spaces: Because noteworthy events in those communities were never documented, the accounts of displaced neighborhoods of color confronting eminent domain are rare. Historic preservation in urban communities of color is challenging, to say the least. Since 2011 in El Paso, we have witnessed the city's effort to remove and demolish schools, artwork, and buildings important to the Mexican American community in quick succession, as if it were open season on Mexican and Chicana/o symbols.[23]

On June 2, 2011, attorney Raymundo Eli Rojas, a Lincoln Park supporter, made a second request under the Texas Public Information Act to the Office of the City Clerk, city of El Paso, for records to be opened to provide the following information:

> 1) any results of asbestos analysis performed after October 1, 2006 of the Lincoln Cultural Center located 4001 Durazno which are not in the Sun City Analytical Inc.'s October 2006 Microbial Investigation;
> 2) any results of rat infestation investigation or analysis performed after October 1, 2006 of the Lincoln Cultural Center located 4001 Durazno which are not in the Sun City Analytical Inc.'s October 2006 Microbial Investigation;
> 3) any interdepartmental memorandum, communication, etc. regarding the demolition of the Lincoln Cultural Arts Center located at 4001 Durazno that were written after October 1, 2006;
> 4) any written communication between Environmental Services, or other entity within the City of El Paso, with the Texas Department of Transportation regarding the Lincoln Cultural Arts Center located at 4001 Durazno.[24]

The documents revealed that Wilson had directed staff to demolish the building with no public input.

At the El Paso City Council Meeting on October 18, 2011, Debbie Hamlyn, the former director of Quality of Life for the city of El Paso, presented her argument for the demolition of Lincoln Center. Among other things she argued that the district had lost so much population that the neighborhood no longer warranted a center.[25] LPCC stated that, aside from its low-income residents, that part of the city had a large undocumented population that has historically been undercounted by the US Census and that there were over six thousand young people in the immediate area of Lincoln Center. The allegations by the city were designed to erase the community, but upon learning more of the area's history, they have ended up doing quite the opposite.

At Annual Lincoln Park Day on September 25, 2011, thousands gathered to celebrate the park and unveil a new mural by Gaytán. Lincoln Park Day is one of the largest regional classic-car shows in the city, averaging over five thousand attendees. At the same time, under the threat of Lincoln Center's demolition, LPCC initiated the Save Lincoln Park campaign by hosting a press conference with attendees and Lincoln residents. On that day LPCC also collected more than five hundred signatures for a petition to preserve Lincoln Center, addressed to former El Paso Mayor Jonathan Cook. LPCC subsequently presented those letters and petitions to city representatives and Mayor Cook directly.

Hamlyn stated the building had issues with mold and asbestos due to the floods of 2006; however, as open records requests have proven, the city failed to produce any evidence of mold or asbestos. At a city council meeting, the city representative for District 3, Emma Acosta, introduced a resolution to grant a reprieve to Lincoln Center for six months while the city communicated with the neighborhoods surrounding Lincoln Center regarding its demolition. The resolution passed unanimously. Several LPCC members, supporters, and community residents spoke on behalf of keeping Lincoln Center open and stopping its demolition. LPCC asked for more time to develop plans to save the building, citing that the community was never notified of the closure in 2006. LPCC then obtained a six-month stay and then began working with the city of El Paso and TxDOT to explore viable options to keep the building being demolished and to reopen it.

Former City Representative Emma Acosta scheduled two public meetings for input on Lincoln Center, one for Wednesday, January 11, 2011, and another one for Saturday, January 14, 2011.[26] At the first public input meeting, held at the Silva Health Magnet School on January 11, 2011, six years after the public facility had been closed, former City Engineer R. Alan Shubert stated no asbestos had ever been found in the building. It was the first time city officials admitted they had obviously communicated inaccurate information about asbestos having been the reason to close the building. As the campaign to reopen Lincoln Center has gained traction, other influential figures signed on to support LPCC in their efforts to open the building.

At the city council meeting on October 4, 2011, LPCC learned that the city did not own Lincoln Center and that TxDOT was the rightful owner.[27] At the same meeting, LPCC learned from City Manager Joyce Wilson that Lincoln Center was slated for demolition and the newly renamed and remodeled O'Rourke Recreational Center (formerly the YMCA) at Virginia and Montana Streets (located four miles away) was its replacement. The O'Rourke Center was miles away from Lincoln Center, and it necessitated crossing Interstate 10 and other busy streets. Wilson had shifted funds from Lincoln Center to the O'Rourke Center, named for Pat O'Rourke, Congressman Robert O'Rourke's late father.

In August 2013 LPCC created a fact sheet titled "Help Us Re-open Lincoln Center!" The fact sheet presented the following questions: What is Lincoln Center and why is it important? Who owns the building and who wants it to be reopened? What requirements does TxDOT have for the reopening of Lincoln Center and who supports the reopening of Lincoln Center? In July 2013, at LPCC's invitation, Hector Gutierrez Jr. became part of the effort to save Lincoln Center from demolition. Gutierrez had been the first Hispanic corps commander of the Texas A&M Corp of Cadets. By 2013 he was a retired lobbyist and strategist with extensive political contacts, and he agreed to help LPCC as a private citizen. He had served as a board member of the Medical Center for the Americas and had a thirty-four-year career as a professional lobbyist, which included tenures at El Paso Electric and Hillco Partners. He had also served as Public Affairs Consultant for the city of El Paso, the El Paso Water Utilities Public Service Board, AT&T, Motorola, and the American Heart Association.[28]

Having firsthand knowledge of the key players in El Paso, Gutierrez proved to be a valuable guide for the preservationists to navigate El Paso politics. It is important to note that members of the LPCC and supporting members (students, scholars, and community activists) all served and continue to serve as community volunteers. LPCC is not a tax-exempt organization, and efforts to turn it into a nonprofit have been resisted by core members. The LPCC organizational structure is organized as a neighborhood association.

On July 12, 2013, Dr. Mariana Chew, president of Xicali Engineering in El Paso, conducted a second walk-through of the building, and she authored a report that refuted the center had mold and water damage and was "unsafe."[29] Her findings stated issues with the building could have been easily addressed by the city. That same year Senator Rodríguez's Sustainable Energy Advisory Committee expressed an interest in "greening" the building as an incentive to reopen it.

Engaging Community to Save Lincoln Center: The Do What's Right Rally

There are only two small green signs that designate Lincoln Center: one at the corner of Reynolds and Durazno and another at the corner of Copia and Durazno. Lincoln Park attracts thousands of visitors every year and serves as an economic booster for the city. From these small signs on July 21, 2013, came the idea to create the Do What's Right rally to raise awareness of Lincoln Center's demolition. The campaign produced a T-shirt and held a petition drive in the park's amphitheater.

In summer 2013 TxDOT still required an institution to consider acquiring the building. LPCC approached EPCC to act as an agency to help reopen the center, so on August 12, 2013, EPCC staff conducted a walk-through of Lincoln Center and presented their options to EPCC President William Serrata via a college-produced report titled *Site Evaluation of 4001 Durazno—Lincoln Center by the EPCC Physical Plant, Police Department and Information Technology Department, 8-12-2013*. The draft report featured a brief history of the building and a summary of what it would take to reopen the building.[30]

EPCC estimated the minimum renovation requirements to reopen Lincoln Center would be $2,437,880.00. LPCC sent a letter to TxDOT Chairman Houghton on August 13, 2013, with electronic copies to Senator Rodríguez, state, and city representatives, and Dr. Dennis Bixler Marquez, then director of Chicano Studies at UTEP, asking Houghton to postpone the demolition of

Lincoln Center until a suitable sponsor for the building could be found, but Houghton did not respond.

Earlier on August 12, 2013, a letter was sent to TxDOT Chairperson Houghton verifying wide community support for the reopening of the center and asking him to remove the demolition clause he had placed on anyone wanting to acquire the building. The multiple use agreement (MUA) was a contract that needed to be signed by the governmental entity who agreed to reopen Lincoln Center. Under "Item 7: Termination," the contract specified that the governmental entity who acquired the center had 120 days to make "required repairs and improvements to the facility to bring the facility into compliance" or the agreement would be terminated.[31] It meant if the governmental entity did not make the needed repairs and improvements in two months, the contract would be revoked. In addition, under "Item 10: Restoration of Area," the agreement required the governmental entity "to within (30) days from the date of said notification, clear the area of all facilities, including demolition of any buildings located on the facility and return the facility in a satisfactory condition to the State."[32] Meaning that if the conditions did not work out for the governmental agency, they would have to vacate in a month and pay for the demolition of any buildings or modifications they had made to the building.

In addition, under "Item 15: Use of Right of Way," TxDOT would not relinquish their "right to use such land for highway purposes when it is required for the construction or reconstruction of the traffic facility for which it was acquired, nor shall use of the land under such agreement ever be construed as abandonment by the State of such land acquired for highway purposes, and the State does not purport any interest in the land described herein but merely consents to such use to the extent its authority and title permits."[33] Meaning that the governmental entity could acquire it and make renovations, but the state could reacquire it at any time they needed.

In summary, the agreement would lease the building but not the land. Under these conditions no governmental agency would enter into the agreement, thus LPCC sent a letter to Chairman Houghton stating that their demolition clause was exclusionary and exemplified a form of environmental discrimination that sought to exclude a specific ethnic group because he or the agency deemed them unfit to open a building in a location with a demonstrated need. The letter also asked Houghton, "If you feel this way about a community group wanting to reopen a center, how will that prejudice cloud your judgment when you are making bigger decisions to benefit Texas' Latino/a growing demographic?" Finally, the group asked Chairman Houghton to "Do what's right."[34]

Like most letters sent by LPCC members, this letter also went unanswered, although it was copied to Senator José Rodríguez, Texas District 29; Texas Representative Marisa Márquez, District 77; Texas Representative Joseph Moody, District 78; Texas Representative Joe C. Pickett, District 79; El Paso County Commissioner, Precinct 1, Carlos Leon; El Paso City Representative Emma Acosta, District 3; El Paso City Representative Eddie Holguin, District 6; El Paso City Representative Lilly Limón, District 7; and Dr. Dennis Bixler-Marquez, director, UTEP Chicano Studies Program.

The MUA was viewed as a form of posturing because TxDOT operates a Real Estate Management and Development arm within its agency that seeks "to optimize its real estate portfolio by leasing and selling real property assets no longer needed for highway use and exploring alternative strategies that have synergy with existing highway assets."[35] Meaning that they sell properties they no longer need, as they have done so in other Texas cities, but it doesn't come cheap.[36]

The I-10 Connect Project

On Sunday, January 26, 2014, an article in *El Paso Inc.* titled "Stop Ahead: Rebuilding I-10" by David Crowder mentioned the linking of the Border Highway at the Spaghetti Bowl.[37] This project became known as the I-10 Connect Project. A press conference was held on February 14 on the steps of Lincoln Center to discuss Alternative 8-2 plan proposed by TxDOT, which would have demolished Lincoln Center. Utilizing political theater, LPCC members created large paper hearts and artist Gaytán created a large heart with a knife going through it with TxDOT letters on it.

KFOX news reporter Stacey Welch broke the story of the demolition of Lincoln Center on March 7, 2014, while it was being presented at the MPO meeting, followed by television stations KVIA and KTSM. Lincoln Center was slated for demolition. TXDOT intended to build a super ramp that would connect I-10 to the Border Highway, and they needed the airspace above Lincoln Center. In response to the city's request for a business plan to reopen Lincoln Center, LPCC wrote an eighteen-page document titled *Business Plan: Lincoln Center for Chicano Cultural Arts, Wellness and Community Archive*.[38] Knowing that reopening the center had long met with obstacles and uncertainty, Joseph Ferguson of the EPCC Small Business Development Center reviewed LPCC's business plan and sent a summary of it to Mayor Oscar Leeser and the El Paso City Council on April 14, 2014.

Lincoln Center's Standoff, the City's Historic Injunction, and Occupy Lincoln Center

On the morning of May 20, 2014, LPCC members and Lincoln Park residents attended the city council meeting and urged them to save Lincoln Center from demolition. The same day, at noon, a fence was erected around Lincoln Center and building supporters went head-to-head with the contractors. TxDOT Chairperson Houghton had ordered the contractor to go inside Lincoln Center and begin their demolition assessment of the building. News of the contractor's presence—that men were putting up a fence and securing the site for demolition—was shared via social media.

When protestors arrived, representatives from JMR Construction stated the site was under their control. Several units of the El Paso Police Department, including a white paddy wagon, were present. Senator Rodríguez arrived on-site and began negotiating with TXDOT officials in Austin to defuse the situation, as did former El Paso District Engineer Bob Bielek.

That evening LPCC began an occupy-style tent city to maintain a presence at Lincoln Park in front of Lincoln Center. The next day, on May 21, 2014, LPCC received a call that contractors had driven up with a Bobcat bulldozer on a trailer. Protesters blocked the trucks (see figs. 15 and 16). LPCC member González stated that things were escalating fast. Meanwhile El Paso City Representatives Lily Limón and Eddie Holguin, with support from Mayor Oscar Leeser, were engaged behind the scenes, creating an injunction to stop the demolition. It became a tense situation as activists waited for word on the injunction. Finally, in midmorning, Representative Holguin arrived at Lincoln Center followed by Representative Limón, who arrived with copies of the injunction.

Contractors refused to leave the premises, and they challenged the legality of the injunction until they were called off by TxDOT officials. This activity signaled a new level of engagement by LPCC and Lincoln Center supporters willing to jump in front of bulldozers and create a human chain to deter the demolition of the building. For a week the Lincoln Center tent city operated as a base camp to monitor the situation. LPCC members were joined by citizens who camped out overnight at the park. They also received donations of water, food, and ice to support volunteers braving the elements.

On June 1, 2014, Corinne Chacon and LPCC President Hector González were interviewed by KVIA news journalist Maria García about Lincoln Center and a poster created by a local artist that featured an image of Chairperson Houghton with the words "Houghton Hates Raza." LPCC members had shared the image on social media, and García asked Chacón and González why members were engaged in "hate" rhetoric. LPCC member Hector González attended the June 26, 2014, Texas Transportation Commission meeting in Pasadena (part of metro Houston), Texas, where Lincoln Center was discussed as item number 6 on the agenda. The meeting was also attended by State Senator Rodríguez, State Representative Joe Pickett, and El Paso Deputy City Manager Sean McGlynn. González attended the meeting to follow up on the city council's reaffirmation of LPCC acquiring Lincoln Center to reopen it. In prior discussions with TxDOT, LPCC stated they would acquire and reopen it if they were allowed to do so.

González presented a brief historical account of the building and its significance to the El Paso community. Senator Rodríguez and State Representative Joe Pickett spoke about the interaction of TXDOT with the community and the lack of communication and transparency on their proposed plans, which would result in the demolition of the center.[39] Up to that point, none of the commissioners other than Ted Houghton and Lt. General Weber seemed to have heard of the issues associated with Lincoln Center. Before the agenda item was discussed, in a last-minute decision, Chair Houghton asked for a five-minute bathroom break, and he informed the public and other elected officials they could leave the building. Over half of the audience left.

Those are some of the tactics elected officials engage in to dissuade public comment. At that meeting several individuals came to speak in support of reopening Lincoln Center, but public comments were moved to later in the meeting. Of course, working people or parents cannot spend an entire day waiting to speak, and they often leave before public commenting is opened.

LPCC members again appeared in front of the Texas Transportation Committee on October 30, 2014, at their meeting in El Paso at UTEP. Once again LPCC asked TxDOT to present testimony about the importance of Lincoln to the El Paso community and how the neighborhood would be affected by its demolition. The commissioners scheduled LPCC members to speak close to the end of the meeting, even though Senator Rodríguez had requested to speak at the beginning to allow community members and others to have their voices heard. By the time LPCC members spoke, most of the attendees had already left the meeting.[40] Commissioners were instructed not to interact with the public on the issue of Lincoln Center, even though Lt. Gen. Weber broke the open meetings rule and answered LPCC even after being instructed to refrain.

Dr. Max Grossman received an e-mail from Sheila Casper from the National Trust for Historic Preservation on July 3, 2014. Casper was looking for historically significant Latino sites that could benefit from assistance from the National Trust. Grossman contacted LPCC members and so began a relationship with the National Trust for Historic Preservation to look at ways of saving and refurbishing the center.[41] The National Trust for Historic Preservation in Washington, DC, identified Lincoln Center as a historically significant Latino site that could benefit from their advocacy and assistance.

In summer 2011, by teaming up with State Senator Jose Rodríguez and other advocates, LPCC members transformed themselves into preservationists in their pursuit of reopening a building. The story of saving Lincoln Center as an urban space has power. Its power is derived from the activism of individuals gathered to resurrect the vitality of the area. It is an issue of claiming culture and sustaining it versus seeing it as disposable. The topic of preservation is now a global issue because of the internet; it forms a new dialogue. According to artist Celia Muñoz, "There is a necessity for a culture to exist." Muñoz went on to state that "urbanization is not only a national issue but also an international dialogue; cities need activists to remind them who are they and what they are—big issue is displacement; it is living history."[42] Ms. Muñoz's statement suggests that activists are usually at the forefront of addressing important issues regarding the changing nature of neighborhoods and when cities overreach and violate people's lives.

Just as highway building became a political act, the painting of murals on the highway pillars at Lincoln Center was an act to affirm identity and resistance in twenty-first century El Paso. In addition, the murals on the pillars supporting the highway exchange above Lincoln Park mirror San Diego's Chicano Park, which also hosts murals painted on its freeway columns.

Lincoln Park's monumental murals are tied to the Chicano mural movement, which has been the subject of significant scholarly evaluation and research in the fields of ethnic history and fine art. Since 1981 regional artists have painted murals on the pillars (pylons) under the Spaghetti Bowl at Lincoln Center. These monumental works are significant examples of mural production in El Paso. The collection of murals at Lincoln Park are significant because they are painted by Mexican American and Chicano artists who are tied to Mexican and Mexican American historical narratives and the Chicano/a mural movement.

Artistic Murals of Identity and Symbols of Resistance

El Paso's murals are products of a unique border culture. El Paso and Ciudad Juárez are two cities in two nations linked via ancestral and historical roots but separated by manufactured barriers—bridges, highways, ramps holding up highways, a concrete-lined river, and immigration check points.

In 1999 Chicano Artist Carlos Callejo, who years earlier (1993–95) had been commissioned to paint a large mural titled *Our History* in the El Paso County Courthouse in downtown El Paso, proposed and completed a mural project for Lincoln Park in conjunction with the Private Industry Council

(PIC). He proposed to work with eighty students from various school districts. The mural project, titled the Spaghetti Bowl Project, also produced a video on the creation of the murals and sponsored art classes at Lincoln Center. The project hired five artists to work with students and community members to produce the murals. The artists included Cesar Inostroza, Fabian Arraiza, Steve Salazar, and two women who were not artists but youth supervisors.

Inostroza and Callejo continued painting murals after funding for the project ended. Callejo said each row of columns was dedicated to various themes: one row was titled "Memorial Walk" and was dedicated to historical figures such as César Chávez, Ruben Salazar, and Martin Luther King, and another row was dedicated to the "Natural Elements," such as Mother Earth and Father Sun. Callejo, who painted approximately thirty-five murals at Lincoln Park, organized a community group that included individuals from the neighborhood to approve the themes for the murals.[43]

Likewise, El Paso's murals cannot abandon the regionalism of imagery common to the border area. In El Paso's Chicano murals, design motifs were taken from the legacy of the twentieth-century mural movement of Mexico and its masters—Diego Rivera, David Alfaro Siqueiros, and José Clemente Orozco. As in Chicano murals painted elsewhere in the United States, the themes draw upon ancient pre-Colombian cultures—Mayan, Aztec, and Toltec—not as icons of the past but as celebrations and metaphors to the richness of contemporary Chicano culture.

Before Callejo's and Gaytán's murals were torn down in 2019, there were approximately fifty-six murals in existence at Lincoln Park. In addition to the outdoor murals, there are three indoor works painted under the supervision of Carlos Enrique Florés, who attended La Academia de San Carlos in Mexico City in the middle to late 1960s.

Art of Protest and Community Mobilization

The creation of Occupy Lincoln Center created an upsurge of creativity, protest art, and community mobilization. The fence surrounding the building put up by contractors became an art piece as activists and community members decorated it with motifs, altars, and symbols of Chicano identity and resistance. In addition, local artists created images of dissent against individuals like TxDOT Chairperson Houghton who were seen as the proponents to tear down Lincoln Center.

Even El Paso artist Hal Marcus recorded a two-minute video on saving Lincoln Center titled "El Paso's Lincoln Center Will Be Saved 10 Reasons by Hal M.," on May 26, 2014. He stated that he had revelations on saving Lincoln Center, and he presented ten reasons why the center needed to be saved. His reason number ten was, "To save Lincoln Center is the right thing to do."[44]

Using Facebook, Twitter, and Change.org, LPCC educated the community about the importance of Lincoln Center, the community, and the need to preserve an important building in a historically significant community. Social media, in conjunction with broadcast media, created project recognition and critical mass about the issue. The internet played a significant role in democratizing the issues in the preservation of Lincoln Center. People could then relate to it, and they could form opinions either for or against the demolition. Ongoing programming helped humanize the building, but social media transformed it into an emotional issue, so much so that supporters were willing to jump in front of bulldozers.

Lincoln Park Day

Beginning in 2005 Lincoln Park has been used by the Chicano community, the Latin Pride Car Club, and the Lincoln Park Conservation Committee to present the annual Lincoln Park Day. In 2010 the LPCC, in partnership with Lincoln Park residents, became a city-recognized neighborhood association and Partner-in-Parks. A Partner-in-Parks designation allowed LPCC to present quarterly social and cultural events. The park features an amphitheater where performers presented dances and activities associated with the annual Lincoln Park Days every third weekend of September. In addition to the September event, LPCC also presented César Chávez Day, the annual Día de la Virgen Guadalupe, and the annual Day of the Dead celebration.

LPCC sponsors four events at Lincoln Park a year, including the Annual Lincoln Park Day, attended by thousands of people. A majority of the members of LPCC are members of various classic car clubs in El Paso who were involved in advocacy and preservation of the Lincoln Park community and Lincoln Center. And in contrast to other Parks and Recreation Centers, which focus on sports, Lincoln Center has remained the only cultural arts center in a Mexican American city up to the opening of the Mexican American Cultural Center in March 2025; it therefore serves as El Paso's Chicano cultural center. Recreational facilities include a twenty-three-acre landscaped park, basketball and handball courts, a children's playground, an amphitheater with a raised stage area, picnic areas and park benches, a flag area, and a 120-car parking lot. Lincoln Park is accessible from Interstate 10 and highway access roads and neighborhood streets.

The monumental murals painted under the pylons at Lincoln Park are less than fifty years old, and they have achieved significance as examples of superior examples of mural production in El Paso.[45] The murals are significant in the state because they were created by Mexican American and Chicano artists that are tied to El Paso's mural movement, as well as artists who exhibited in the Lincoln Center Gallery and, most recently, via the creation of annual cultural festivals that occur on the grounds of Lincoln Park.

Efforts to Save Lincoln Park's First Murals

On June 30, 2018, TxDOT District Manager Robert Bielek reported that due to the I-10 Connect Project, Lincoln Center would be saved, but that Lincoln Park's first murals painted under the supervision of Carlos Callejo, and several murals by Gabriel Gaytán, would be torn down to make way for a wider ramp to accommodate traffic as part of the I-10 Connect Project, but Felipe Adame's 1981 mural titled *Virgen de Guadalupe with Roses*, would be saved.

The LPCC, which was the caretaker of Lincoln Park, met with Bielek and TxDOT staff on June 3, 2018, and presented three options for the murals to be affected because of the I-10 Connect project.[46] The news that the building was not going to be torn down but that sixteen murals were going to be removed created a win-lose scenario, which according to TxDOT was unavoidable to save the building and the amphitheater. LPCC has been working on saving Lincoln Center from demolition since 2009.

LPCC presented TxDOT with several options to save the murals and/or to compensate artists for the loss of their art, which consisted of Options 1, 2 and 3. Option 1 saw the redesign of overpasses over Lincoln Park to save all murals. The committee recommended that overpasses be tall enough to save all the column murals, and that new columns should be placed, ideally between the existing ones for the murals to be seen and not blocked by the new columns. LPCC saw this option as most cost prohibitive, and it would require other changes in TXDOT's existing plans to reroute traffic from I-10. Option 1 would have been the best option for the community to preserve Lincoln Park's historic murals.

Hector González, president of LPCC, stated the removal of the murals was not going to sit well with the community. Bielek stated that Option 1 was impossible unless LPCC wanted to get rid of the building (Lincoln Center). Decisions were based on budgetary considerations due to the MPO being in a lapse-grace period with the Horizon 2040 plan being frozen, meaning that Bielek could not change the scope of the present I-10 Connect or its funding.

Bielek explained that the span length determined where the new columns would be placed, and that elevation was more of a function. He stated that one part of the determining factors was the economy or what funds were in place for the project. He stated uniformity was one thing that required the elements to be tied together.

Option 2 would also have seen the option of saving the murals by just removing top portions of the pillars and saving all the murals, but LPCC felt the integrity of these art works would be destroyed. Most of the murals would be chopped off at the top, and in some instances the heads of mural figures would be removed. In addition, TxDOT's plan was to place new pillars six feet in front of the murals, and they would have obstructed the view of the saved pillars. Murals are meant to be seen from a distance, not from several feet away, which this plan promoted. This plan would also have added additional costs.

Concerning Option 2, Bielek agreed to explore it at a planned meeting but unfortunately, the murals would also be compromised with this option. He also stated that cutting the tops of the murals would not be feasible and that the contractor would not be able to do it. Bielek stated Option 2 was impractical because we would also lose the integrity of the murals. Option 2 was also too costly and would not have worked.

Option 3 would see the removal of the pillars entirely, and in exchange TxDOT and LPCC would jointly apply to the National Registry of Historic Places to make Lincoln Park and Lincoln Center historic sites. In addition, LPCC recommended that TxDOT compensate artists to repaint new murals on new pillars for $20,000 per mural, times sixteen, for a total of $340,000, plus an equal amount of $340,000 compensation for the loss of these historical works and the community impact (funds to restore remaining murals and make other improvements in the park, such as improving the amphitheater and park grounds). The total amount was $680,000. LPCC requested the preservation of the late Felipe Adame's 1981 mural *The Virgen de Guadalupe with Roses* (located at the corner of Durazno and Uva Streets). If this option were pursued, LPCC also suggested TxDOT fund the documentation for historical purposes of murals to be removed. The documentation of these murals would be placed in a local archive, such as the UTEP Special Collections or the City of El Paso Municipal Archives or at the El Paso Public Library. In the meeting with LPCC, Bielek stated Option 3 was the more practical. He said there would be federal grants available to make Option 3 a reality. He stated that once the MPO got out of its lapse, TxDOT could apply for Transportation Security Administration (TSA) funds. Bielek encouraged González to meet with TxDOT officials and see what columns would be affected and which ones could be used. He also stated that a municipality or city and country agencies could also apply for TSA funds. LPCC members asked if TxDOT would support putting the building and the park on the National Register since TxDOT (Maryellen Russo, Alexis Reynolds, and Rebecca Wallisch) had used a 2015 draft application submitted to Greg Smith, Texas State Historic Preservation Office (SHPO), extensively for the June 2018 TxDOT Historical Resources Survey Report for the 1-10 Connect Project.[47]

Concerning the documentation of the murals to be affected, Bielek agreed to send out TxDOT photographers to document them. He stated TxDOT would work with the city and the MPO to explore Transportation Alternatives Program funding. An application for funding would have to come from a local sponsor who would apply for the funds and that it could be either the city or county. TxDOT staff stated that the city or the county would also apply for a Pedestrian Enhancement Project, which could include artwork as part of the plan.

At a meeting in 2018, Bielek stated it might be too early to start working on a Pedestrian Enhancement Project because those grants had a shelf life and the construction in Lincoln Center would begin December 2018 (constructed started February 2019) and would not be completed until May 2021 (construction was set to a year and a half with incentives for early completion). In a meeting with LPCC, Bielek stated that he had "no use for Lincoln Center" and the TxDOT Real Estate Division would have the final say on what happened to the building (he stated TxDOT purchased the building in 1969 for $500,000, but it was closer to $200,000). He also stated that the TxDOT Right-of-Way Division controlled the park. He said the previous memorandum of understanding had placed the onus on the city of El Paso for maintaining the park.

González stated that saving the first mural painted in Lincoln Park, the Virgen de Guadalupe mural by Felipe Adame, was paramount for the community and that the mural needed to be protected at all costs. Fortunately, Adame did not paint his mural all the way to the top, and if the top was cut off, the mural could still be saved.

Bielek then stated he could not speak for the MPO, but he assured LPCC members that he could give them a letter guaranteeing that the district office would help the organization apply for a Pedestrian Enhancement Project grant with art components to satisfy the requirements of Option 3, and/or explore other means to bring the murals to fruition.

During the construction first Lincoln Center would be surrounded by fencing by the contractor to protect the building, then the monumental task of cutting the eight columns would begin. LPCC asked to supervise the work on Felipe Adame's Virgen de Guadalupe mural, where the community would be present.

The removal of Callejo's murals was not without controversy in the community and with the artist who divides his time between El Paso and Los Angeles. Although the images belong to the artists, the columns, the surfaces where the murals have been painted, and the location of the murals belong to TxDOT because of right-of-way, designated with the creation El Paso's freeway system. Such is the issue regarding murals, which tend to be ephemeral works. LPCC and the community had planned to pursue the promises made by TxDOT to restore Lincoln Park and its historical murals back to their original condition before the I-10 Connect Project.

Unfortunately, on Friday, May 10, 2019, Bielek abruptly resigned as district engineer from the TxDOT District 29 office just "as TxDOT had just began their $96 million 1-10 Connect project at the Spaghetti Bowl in April and [had] other on-going projects that [totaled] over a $1 billion (about $3.1 per person in the US)." Projects included "the Go 10 and Border West Expressway near Downtown," the article stated.[48] Tomas Trevino, P. E., was appointed new El Paso district engineer effective July 10, 2019.[49] With the departure of Bielek, LPCC had to reinitiate the process of having TxDOT make good on what they had promised.

Unfortunately, Bielek did not leave any records, notes, or written agreements of the negotiation between the LPCC members and his office. The new District Engineer Trevino said he did not know anything about the arrangements Bielek had agreed to, which were to replace the murals that had been torn down or to make good on the beautification projects, but he agreed to meet and discuss it. To his credit, Trevino initiated a series of walk-throughs with his staff in the park during the construction with LPCC members to discuss enhancement projects.

Conclusion

Righting the Wrongs of the Past

Highway building follows the chronology of historical events, but so does opposition to it. This work originated from the need to uncover the history of Lincoln Park School in the hope that the building would be spared from bulldozers. This work is also a history of dealings between businesspeople who profited in a period when Mexican Americans and African Americans were not at those tables. It was a period when racial covenants were plentiful in El Paso's neighborhoods and minorities could only live in specific neighborhoods.

Even though creating El Paso's early freeways like Interstate 10 displaced families and communities, it also offered opportunities for landlocked people to move to other locations in El Paso. African Americans moved to the East Side, as did Mexican Americans. Families would only be able to move to homes within the amounts they were given for their homes, which ranged from $4,000 to $6,000 and often lesser amounts.

On a cool summer day on June 12, 2019, at 7:00 p.m., the Sunset Heights Neighborhood Association had their monthly meeting at the El Paso County Historical Society in the El Paso County Historical Society's Burges House at 603 West Yandell Street. Sunset Heights is a historic community north of Downtown

El Paso. The well-attended neighborhood association meeting included artists, writers, and journalists who came to hear about new developments in the creation of the project that has been referred to as "The Trench."[1] Former Texas State Representative Lina Ortega, who supported the creation of a proposed deck park, spoke about the merits of the project and why she was supporting it.

The project was twofold: Not only was its intent to place a cover over Interstate 10 from Santa Fe to Campbell Streets, but it also proposed the widening of Interstate 10 from downtown El Paso to Raynor Streets. The plan would place a deck over downtown El Paso and create a usable space like a park or an arena. At the time TxDOT did not have the funds to pay for the Trench, but the state agency has initiated community meetings to discuss the project and Texas Governor Greg Abbott had recently appointed former City Manager Joyce Wilson, who oversaw the demolition of the El Paso City Hall building and the creation of the Southwest Baseball Park, to the Camino Real Mobility Authority, an agency that will oversee the creation of the Trench.[2] It is all a behind-the-scenes moving game of chess to cash in on transportation funding being touted as a once-in-a-lifetime opportunity, but for whom?

According to a May 30, 2019, *El Paso Times* article by Vic Kolenc, "Several city blocks of buildings would be demolished, including the El Paso Holocaust Museum, gateways would be constructed and parks possibly added above Interstate 10 under concepts now being studied for the remake of the highway's 50-year-old Downtown segment."[3] The Sunset Heights Neighborhood Association hired a well-respected planning firm, Creosote Collaborative, to help them understand the Trench project and its ramifications on their neighborhood. Aside from several articles that have appeared in local newspapers and in broadcast media and community presentations that have been presented, TxDOT has not shared how the rest of the city will be affected.

In 2021 with the bipartisan infrastructure deal (Infrastructure Investment and Jobs Act), city planners began crafting equity statements wanting to right the wrongs created by the 1959 Interstate Highway Act, which plowed through minority communities throughout the United States. It's not that during the Covid-19 politicians and planners suddenly had epiphanies; it is because the imminent passage of the bipartisan infrastructure deal would provide over $110 billion for transportation projects.

Touted as a "once-in-a-generation" infrastructure bill, the passage of the Bipartisan Infrastructure Deal "reauthorize[d] surface transportation programs for five years and invest[ed] $110 billion in additional funding to repair our roads and bridges and support major, transformational projects."[4] In an interview on Amanpour and Company on March 19, 2012, Transportation Secretary Pete Buttigeig stated that "public transportation is part of public health." He also stated that cities "ought to have funding that is specifically committed to reversing some of the harms in the past."[5]

In El Paso TxDOT was planning for Phase 1 of the Reimagine I-10 Project. In 2021 the ideas for downtown El Paso and the creation of a deck park atop the freeway were just concepts and no schematics existed.[6] Today there is a fast-track move to get all the pieces in place, including a Section 106 assessment. The Paso del Norte Health Foundation created a website to convince El Pasoans that they need a deck park like other cities. It is no coincidence that the deck park would be at the bottom of the new West Star Building in downtown El Paso. In 2023 it is no longer being called a deck park, it has now been rebranded as a downtown deck plaza, which will have offices, recreational areas, performance areas and even a dog park.

In El Paso Mexican Americans, African Americans, and whites were displaced by highway building. Today, over sixty years later, politicians, chamber of commerce members, city planners, medical centers, and organizations like the Paso del Norte Health Foundation are preparing for the next opportunities. Highway projects command large budgets from taxpayer funds. Who we elect is key to these processes. Unfortunately, many elected officials lack the training and expertise on highway and preservation projects. It behooves us to elect individuals who will side with their constituents and not be lured by big-money elites to vote in a certain way in bodies like the MPOs that decide on future highway projects.

Another aspect of highway building was that these freeway projects were built in a period when there was little public information on the danger of emissions and the future efforts for a green environment. And if freeways, a trench, and a deck park were not enough, Congresswoman Veronica

Escobar, with support from the federal Infrastructure Bill, the governor of New Mexico, and Mexican leaders, is planning to rebuild the BOTA, as it is called locally. Not everyone was happy about rebuilding it. A local community group, Familias Unidas, who has long fought back against environmental oppression in the Chamizal area, held a press conference at the Chamizal National Memorial on April 4, 2022, right before Congresswoman Escobar and officials presented "three 'viable alternatives' to 'modernize' the aging Port of Entry." Unfortunately, as Dr. Kathy Staudt stated in a Facebook post in the group El Paso Street Coalition, all these projects will "take public spaces, undermine residents' quality of life & health for more trucks to cross the border at the Bridge of the Americas." "In the Chamizal Agreement of the early 1960s, the free bridge [El Puente Libre] was built for the people, not for corporations, Clean Air is a Right!" Dr. Staudt added.[7]

How can we right the wrongs of the past? El Paso's early roads and freeways—from the Bankhead in 1915, to Paisano in 1948, to Interstate 10, Highway 54, Highway 110, and the Border Highway—plowed through El Paso and removed hundreds of homes, businesses, and churches and displaced and uprooted countless lives. In the 1960s the El Paso Chamber of Commerce's Highway Committee was the booster of El Paso's early freeways. Again, the chamber and downtown supporters can already taste the creation of the deck park to supposedly "right the wrongs of the past." There needs to be a wider community dialogue and input on all the options being presented.

Citizens organized themselves long before these projects developed. Organizations like the Sunset Heights Neighborhood Association, the Community First Coalition, and efforts like El Paso Street Coalition hope to make a difference by arming themselves with facts and information. Regardless, there is one bright spot on the horizon. The Mexican American Cultural Institute was able to lease Lincoln Center from TxDOT in January 2023. The nonprofit hopes to reopen the center as a Mexican Cultural Arts and History Center in the future, while the city of El Paso opened its own Mexican American Cultural Center.

Appendix
El Paso's Highway Builders

District Engineer Joe M. Battle, Builder of El Paso's First Highways

Journalist María Cortés, in an *El Paso Times* article, reported "Battle served 45 years in various jobs with the state highway department . . . and began working in El Paso as a member of a surveying crew in 1937 while a student at the University of Texas at Austin."[1] Battle constructed Houston's expressway system as part of the Houston Urban Project.[2] He joined the El Paso District Office as assistant director, succeeding E. W. Mars upon Mars's retirement as district engineer for West Texas. It was reported that initially Battle was not happy with his promotion since he felt he was leaving his home in Houston.[3]

Battle arrived in El Paso on September 1, 1963, eight months after the US government settled the Chamizal Dispute on January 14, 1963. In addition, Interstate 10 was making its way to El Paso and Loop 375 had already been planned.[4] What made this assignment different for Battle was that in addition to increased highway funding for El Paso, there was also the need to create infrastructure for the growing regional commercial twin plant industry.

Battle inherited the challenging task of facilitating the creation of El Paso's early highways, the creation of farm-to-market roads, working with state and federal governmental agencies on both sides of the US-Mexico border, the building of ports of entry after the Chamizal Settlement, and the building of the Trans-Mountain Road, initiated by Judge Woodrow "Woody" Bean. Battle was the right person at the right time for El Paso. He arrived as an experienced engineer at a time of vast growth. Also, many projects Battle worked on needed to be built in fifteen years, as mandated by Congress. Joe Battle stepped down as district engineer on October 3, 1991, after twenty-eight years of service, and the city council voted to name part of Loop 375, which was under construction at the time, after him. Battle died March 19, 2004, fifteen days before his 89th birthday.[5]

Tom Diamond, Texas Highway Department's First Right-of-Way Engineer for El Paso

The late Tom Diamond, who received a degree in civil engineering from Stanford in 1956 and a law degree from Baylor University in 1957, arrived in El Paso in late 1958 to open the city's first right-of-way bureau to acquire private property for the construction of Interstate 10.[6] At the time the right-of-way process was handled by the cities.[7] Diamond stated that "sometime in 1956 or 1957, the state law was changed to permit the state to acquire right-of-way instead of counties and cities." Diamond said that less than 10 percent of the acquisitions were comprised of eminent domain efforts.[8]

When Diamond arrived in El Paso as the city's first right-of-way agent, he recalls that "there was no right-of-way acquisition program whatsoever by the city of El Paso." His recollection was also that "the county of El Paso had been purchasing right-of-way for Interstate 10 both east and north of the city of El Paso."[9] Diamond's right-of-way department eventually grew to six or seven employees. Diamond recalls that Battle was "a very fine engineer" and had "a very keen intellect; very professional in his approach to everything; and a good person to get along with in general; he didn't anger easily; [he was] very calm, methodical and he did a fine job."[10]

Diamond had been a part of the El Paso Chamber of Commerce's Highway Council, serving as chairperson of the council for ten years. When asked as to who decided on the path of Interstate 10, Diamond stated the decision had been made before he arrived in El Paso in 1958. He said he had spoken to an engineer and surveyor named Corbett, who had been responsible for surveying the route.[11] Diamond said Corbett had told him that the location of the freeway had been determined by the north line of the Ysleta and Socorro Grants.

Diamond remembered the opening of the freeway when city officials held a formal ceremony on Interstate 10 on December 1968. He said everyone was parked on I-10, and when he cut the ribbon, he took the microphone and yelled, "Gentlemen, that concludes our program, start your engines!" He laughed and said they all looked at him, then went and started their engines and the freeway was officially opened.[12]

Jonathan Ray Cunningham, El Paso Director of Planning, 1958–1978

Jonathan Ray Cunningham was another important figure in the creation of El Paso's freeways. Joan Cunningham-Estrada said her late father's parents moved from Indiana.[13] Cunningham's grandfather had been a professor at New Mexico State University, and his mother had moved to the Southwest to recover from tuberculosis. Cunningham grew up in Las Cruces, New Mexico.

Cunningham-Estrada said that her father earned a bachelor of arts degree in political science from New Mexico State University and a master of arts degree also in political science from the University of New Mexico, then took courses at UCLA until he was drafted for World War II.[14] She said he served in

World War II as a historical officer and wrote the history of the Trans-Pacific Air Command. After World War II he attended classes at the University of Chicago. He taught at Wittman College in Washington State for several years and eventually became a city planner in the surrounding community of Walla Walla, later becoming a city planner in Spokane, Washington.

In Spokane he was recruited by El Paso Mayor Raymond Telles to come and work as a city planner. The Cunningham family moved to El Paso in 1958 "when it was a little town."[15] Cunningham was a learned person at a time of momentous change in El Paso. He served as director of planning for the city of El Paso from 1958 to 1978, a period that included the decade of highway building from 1960 to 1970.[16]

He was also involved in the historic Chamizal Settlement, which would have been a challenge for any city planner because it included working with two nations, two presidents, and, over the years, a plethora of mayors, politicians, communities, and special interests.[17] Cunningham was also a diplomat, having survived the administrations of nine El Paso mayors and varied personalities who made up the El Paso Chamber of Commerce Highway Council, as well as numerous politicians who were often larger than life.[18]

On November 3, 1977, Mr. and Mrs. Jon Cunningham were invited by First Lady Rosalynn Carter to attend a ceremony at the Chamizal National Memorial in El Paso, Texas at 3:00 p.m. to thank individuals who had worked to create the memorial.[19] The meeting was also attended by Carmen Lopez-Portillo, wife of Mexican President José Lopéz Portillo, who, along with President Lyndon B. Johnson, had settled and signed the Chamizal Settlement in 1963. Other dignitaries in attendance included US Representative Richard White, California Governor Jerry Brown, New Mexico Governor Jerry Apodaca, Alderman Dan Ponder and his wife, Executive Assistant Jim Kirby, County Judge Udell Moore, and former County Judge Travis Johnson. Before proceeding to the BOTA, thousands of spectators waited along the bridge and on the Mexican side as the two first ladies passed by Zavala School, which had been spared by the creation of Highway 110. At Zavala School, "Several hundred school children filled the playground to cheer and wave."[20]

Both first ladies met in El Paso and Ciudad Juárez to "commemorate the signing of the Chamizal Treaty and the 10th anniversary of the park."[21] Cunningham-Estrada stated her late father encouraged the federal government to create a national park to commemorate the community, but his job was not all pleasurable. When asked how her father dealt with the multiple interests in city planning, Cunningham-Estrada said he was very "adaptable and he learned how to work with power without being offensive, but I remember him coming home and being so tired from the Planning Commission meetings and sometimes being so frustrated from having to work with the special interests." She described her late father as an idealist and a very practical man who had to "play the party game and go step-by-step and not buck the system too much." Cunningham-Estrada said, "You still had to have great integrity and lead and that is what he always did."[22]

Jonathan Cunningham had a bird's-eye view of city planning and the politics of highway building. In 1969 he hired Nestor Valencia as a planning technician. DURING his tenure as director of city planning, Cunningham groomed Valencia to succeed him after his retirement in 1978. Valencia followed Cunningham as director of planning, research, and development in 1979 until his retirement in 1991.[23] But by 1979 all El Paso's early freeways had been built.

Cunningham played a significant role in the creation of El Paso's freeways, but as a trained planner, he had to work within certain limits, knowing the city's plans would disrupt people's lives and the fabric of entire communities. Urban historian David Díaz has argued that "planning [has] assumed a structural role by ignoring the gross deterioration of the urban condition of minority communities."[24] Yet in many ways, Cunningham's hands were tied as a public employee; he could not counter El Paso's business elites, politicians, and policy makers. As Cunningham-Estrada stated in her interview, her father had to learn to adapt and had to learn "how to work with power" without disagreeing with his superiors and other special interests.[25] Yet the relocation of citizens in urban or road projects was a constant issue from 1960 to 1970, in El Paso and throughout the country.

Manuel F. Aguilera, the Spaghetti Bowl, and Highway 54

The late Manuel F. Aguilera, who rose to deputy director at TxDOT, was born five blocks from the Lincoln Park community in his grandmother's house at 77 Boone Street at a time when it was common to be born at home with the assistance of midwives.[26] Aguilera was a Fronterizo (a border person), born in Mexico but raised in both countries, who later became an engineer and helped build El Paso's freeways. His father was from Guadalupe, Mexico, across from Fabens, Texas. His mother was from El Paso. Raised in Guadalupe until he was 6 years old, Aguilera then came to live with his grandmother so he could attend school in El Paso. On weekends his parents would pick him up, and they would return to Guadalupe. He attended Burleson School, which later became Jefferson High School. After Burleson closed, he attended Zavala School in the Latta's Woodlawn Addition, a neighborhood south of the Lincoln Park community, which eventually became Highway 110.[27]

In 1964 Aguilera graduated from Texas Western College (now UTEP) with an engineering degree. Numerous Chicano engineering students graduated from TWC. After graduation he was drafted to Vietnam. He stated, "I was very fortunate because I already had my degree and went to Fort Poke,

Louisiana for basic training and was the only one in my company who didn't go to Vietnam."[28] He was sent to Arkansas to work in engineering. After two years in the military, he was offered a job in Seattle with Boeing. In Seattle, he said they were building the first 747s but after being there for a week and not seeing the sun, he decided to return to El Paso for a position at White Sands Missile Range. Two months before he was to begin his new position at White Sands, he went to visit one of his professors at Texas Western College who told him the Highway Department was hiring engineers.

Aguilera said that by 1967 the design for Interstate 10 had been drawn up. His engineering professor called the El Paso Office of the Highway Department, then located on Clark Street (currently the building for the Parks Services) and scheduled an interview for his protégé. The next day Aguilera interviewed and was hired. He said the pay at the Highway Department was $20 less per month than the position at White Sands, but he did not mind it because he would not have to drive there. Aguilera began working on the design of Highway 54 from Interstate 10 to Pershing Street.

According to Aguilera, the creation of Highway 54 included different sets of engineers working with issues of flood control and terrain. Aguilera said Lincoln School became the field office for all the engineers in the construction of Interstate 10, the Highway Interchange, and the North-South Freeway (US 54).[29] He stated, "One of our inspectors, I remember going into the restroom at Lincoln Center, it was a grade school and all they had were toilets for little kids . . . and here is this guy 6' 4" or 6' 3" sitting on a toilet with his legs up here [pointing to his shoulders]."[30] Aguilera said the right-of-way costs were the jurisdiction of the city or the county. The property underneath became a park after the construction of I-10.

When Aguilera began his position with the Highway Department, he said there were many Mexican American engineers who were mostly graduates of Texas Western, but that all the managers were white. Even though he was an engineer, Aguilera said he faced discrimination. He said, "Back in the 1960s, you had to be two times better than the Anglo guy to move up."[31] Aguilera's first position was in the design department when the interchange was already in the final design stage. He said the design portion for the four-level interchange was so massive that it was developed both in El Paso and in Austin, as well as farmed out to engineering companies in San Antonio and Houston.

A 1967 black-and-white photograph shows Aguilera with fellow engineers in front of a maquette of the interchange (locally known as the Spaghetti Bowl) along with a foot-high stack of blue-line plans for its construction (see fig. 10). He said when the project was awarded it had an estimated cost of $19 million dollars, and the job came in at $17 million—the highest price for a highway project at the time.[32]

Aguilera said he worked in the design department of the El Paso office of the Texas Highway Department for eight years, where he was tasked with creating the North-South Freeway. The department maintained a close working relationship with the city of El Paso because they had to purchase the property and clear it, then the engineers would design and build the freeways. For the construction of Highway 54, the city bought entire city blocks of houses from I-10 to Pershing Street. The Highway Department had a staff of workers who worked with the city to acquire properties.

According to Aguilera, one engineer was directed to design the section of US 54 from Pershing to Van Buren, Aguilera developed the part from Van Buren to Fred Wilson, someone else designed the section from Fred Wilson to Hondo Pass, and another from Hondo Pass onward; all teams worked on building the freeway at the same time. He said they developed US 54 with flood control all the way to the Texas–New Mexico line and worked with Fort Bliss because the highway was adjacent to their property, and they had to move some of their flood control dams and rebuild some of them, like Pershing Dam. He said the project required considerable coordination.

Aguilera remembered that in 1968 or 1969 he test-drove a Jaguar he had bought downtown at one hundred miles an hour on Interstate 10, but then the freeway ended at Copia Street. Aguilera said the last section of I-10 to be built went past UTEP going north. He designed the section of I-10 from Mesa to Sunland Park to Executive Center Boulevard and later worked on the design of the Chamizal Freeway (today called the Border Highway). He also designed the pump stations for pumping the water from Cordova Island into the Rio Grande. Aguilera said the design took drainage and terrain into consideration, including the city's responsibilities to maintain the flood control dams. He said Battle made the decision to construct the highways out of reinforced concrete because of its long life. He said reinforced concrete was costly then, but over the decades it has maintained its effectiveness.[33]

Aguilera said one day Battle requested a meeting with him and asked him if he wanted to work as a traffic engineer.[34] Aguilera told him he did not know much about traffic engineering, so Battle sent him to Northwestern University for three months to learn it.[35] From then on he worked as a traffic engineer, then a deputy assistant director in the 1990s. When the TxDOT Laredo district engineer position opened, he said he thought about applying, but because his children were in college, he declined it. Aguilera became one of the first Hispanic engineers to break through the ceiling and become the first Mexican American assistant district deputy director under Battle. He and other Mexican American engineers opened the door for the next generation of transportation engineers. When he retired in 2003, Aguilera received recognition for his work in building El Paso's roads. The county of El Paso named a highway in his honor. Aguilera Highway runs from Tornillo to the new port of entry to Interstate 10.[36] Aguilera passed away on October 24, 2017.

Endnotes

Notes for the Introduction

1. Editorial, "Fifty Candles and Many Highways, 1916–1966," *American Highways,* 1966, p. 13.
2. Marguerite E. Threadgill, "History and Overview of McNary, Texas," *Handbook of Texas Online*, updated July 31, 2020, https://tshaonline.org/handbook/online/articles/hlm49. Threadgill states, "McNary, formerly called Nulo, is at the intersection of Interstate Highway 10 and State Highway 20, two miles from the Rio Grande and twenty-three miles west of Sierra Blanca in southwestern Hudspeth County."
3. "Convict Labor for Road Work," *Monthly Review of the U.S. Bureau of Labor Statistics* 4, no. 4 (1917): 591–95.
4. "Right-of-way," Merriam-Webster, accessed June 28, 2025, https://www.merriam-webster.com/dictionary/right-of-way.
5. Robert D. Bullard, Glenn S. Johnson, Angel O. Torres, "Building Transportation Equity into Smart Growth," in *Highway Robbery: Transportation Racism & New Routes to Equity*, ed. Bullard, Johnson, and Torres (Cambridge, MA: South End Press, 2004), 179.
6. Trish Long, "Presidents Johnson, Mateos Meet in El Paso to End Century Old Boundary Dispute," *El Paso Times*, September 21, 2024.
7. Julian Lim, *Porous Borders: Multiracial Migrations and the Law in the US-Mexico Borderlands* (Chapel Hill: University of North Carolina Press, 2017), 3.
8. Will Guzman, *Civil Rights in the Texas Borderlands: Dr. Lawrence A. Nixon and Black Activism* (Urbana: University of Illinois Press, 2015), 33.
9. Ronald William Lopez, "The Battle for Chavez Ravine: Public Policy and Chicano Community Resistance in Post-War Los Angeles, 1945–1962" (PhD diss., University of California, Berkeley, 1999), 12. Note: For a discussion of housing and housing segregation, see Colin Gordon, *Mapping Decline: St. Louis and the Fate of the American City* (Philadelphia: University of Pennsylvania Press, 2008); Kenneth T. Jackson, *Crabgrass Frontier: The Suburbanization of the United States* (New York: Oxford University Press, 1985); Loren Miller, "The Protest Against Housing Segregation," in "The Negro Protest," special issue, *Annals of the American Academy of Political and Social*

Science 357 (Jan. 1965): 73–79; Stefan L. Brandt, "The City as Liminal Space: Urban Visuality and Aesthetic Experience in Postmodern U.S. Literature and Cinema," *Amerifastudien / American Studies* 54, no. 4 (2009): 553–81; Albert M. Camarillo "Navigating Segregated Life in America's Racial Borderhoods, 1910–1950s," *Journal of American History* 100 (Dec. 2013): 645–63; and Camarillo, "Chicano Urban History: A Study of Compton's Barrio, 1936–1970," *Aztlan: A Journal of Chicano Studies* 2, no. 2, (1971): 79–106.

10. Note: For a discussion of race and urban renewal see Scott Kurashige, "The Many Facets of Brown: Integration in a Multiracial Society," *Journal of American History* 91, no. 1 (June 2004): 56–68; Eric Avila and Mark Rose, "Race, Culture, Politics and Urban Renewal: An Introduction," in "Race, Culture, Politics and Urban Renewal," ed. Eric Avila and Mark Rose, special issue, *Journal of Urban History* 35, no. 3 (March 2009): 335–47; Avila, "Revisiting the Chavez Ravine: Baseball, Urban Renewal and the Gendered Civic Culture of Postwar Los Angeles," in *Velvet Barrios: Popular Culture and Chicana/o Sexualities*, ed. Alicia Gaspar de Alba (New York: Palgrave Macmillan, 2002), 125–39; Avila, "Popular Culture in the Age of White Flight: Film Noir, Disneyland, and the Cold War (Sub)Urban Imaginary," *Journal of Urban History* 31, no. 1 (2004): 3–22; Avila, *Popular Culture in the Age of White Flight: Fear and Fantasy in Suburban Los Angeles* (Berkeley: University of California Press, 2004); Eduardo Contreras, "Voice and Property: Latinos, White Conservatives, and Urban Renewal in 1960s San Francisco," *Western Historical Quarterly* 45, no. 3 (Autumn 2014): 253–76; Robert B. Fairbanks, "The Failure of Urban Renewal in the Southwest: From City Needs to Individual Rights," *Western Historical Quarterly* 37, no. 3 (Autumn 2006): 303–25; Fairbanks, *War on Slums in the Southwest: Public Housing and Slum Clearance in Texas, Arizona, and New Mexico, 1935–1965* (Philadelphia: Temple University Press, 2014); William G. Grigsby, "Housing and Slum Clearance: Elusive Goals," *Annals of the American Academy of Political and Social Science* 352 (Mar. 1964): 107–18; Claude Gruen, "Urban Renewal's Role in the Genesis of Tomorrow's Slums," *Land Economics* 39, no. 3 (Aug. 1963): 285–91; Paul R. Mullins and Lewis C. Jones, "Archaeologies of Race and Urban Poverty: The Politics of Slumming, Engagement, and the Color Line," Historical Archaeology 45, no. 1, (2011): 33–50; Hugh O. Nourse, "The Economics of Urban Renewal," *Land Economics* 42, no. 1 (Feb. 1966): 65–74; James

W. Russell, "Class and Nationality Relations in a Texas Border City: The Case of El Paso," Aztlán: A Journal of Chicano Studies 16, nos. 1–2 (1985): 217–39; Ramón Saldívar, "Lyrical Borders: Modernity, the Nation and Narratives of Chicano Subject Formation," Narrative 1 (Jan. 1993): 36–44; Clarence E. Schermbeck, *Urban Renewal for Texas* (Austin: Institute of Public Affairs, University of Texas, 1957); Frank N. Schubert, *On the Trail of the Buffalo Soldiers: Biographies of African Americans in the US Army, 1866–1917* (Wilmington, DE: Scholarly Resources, 1995); and Robert D. Bullard and Glenn S. Johnson, eds., *Just Transportation: Dismantling Race and Class Barriers to Mobility* (Stony Creek, CT: New Society, 1997).

11. Brian D. Behnken, ed., *The Struggle in Black and Brown: African American and Mexican American Relations During the Civil Rights Era* (Lincoln: University of Nebraska Press, 2011); Gerald Horne, *Black and Brown: African Americans and the Mexican Revolution, 1910–1920* (New York: NYU Press, 2005); Quintard Taylor, *In Search of the Racial Frontier: African Americans in the West, 1528–1990* (New York: W. W. Norton, 1998); and Matthew C. Whitaker, *Race Work: The Rise of Civil Rights in the Urban West* (Lincoln: University of Nebraska Press, 2005).
12. Miguel Juárez, "From Buffalo Soldiers to Redlined Communities: African American Community Building in El Paso's Lincoln Park Neighborhood," in "New Directions in Black Western Studies," special issue, American Studies Journal with American Studies International 58, no. 3 (2019): 107–24
13. George D. Torok, *From the Pass to the Pueblos: El Camino Real de Tierra Adentro National Historic Trail* (Santa Fe, NM: Sunstone Press, 2012), 10. The road was used from its founding in 1598 to the 1810s, until it was replaced by the railroads in the late nineteenth century (11, 13). Torok writes that "400 miles of the original trail lie within the United States today, and stretch from present-day San Elizario, Texas to Santa Fe, New Mexico" (13). Parts of the trail were used for north-south improvements in New Mexico and constituted segments of the first federal highways in the region (35). In El Paso "the road became a branch of US 85 which led from the Mexican border in El Paso, along the Rio Grande, north through the central region of the state, and on to Albuquerque" (35).
14. J. F. Friedkin, United States Commissioner, *A Preliminary Report on United States Problems in Acquisition of Private Properties*

in Connection with the Proposed Chamizal Settlement (International Boundary and Water Commission, (August 2, 1963). Brenda Porras, United States International Boundary and Water Commission (Agency) (2016), request by the author, released under the Freedom of Information Act (FOIA) request no. 2017-04.

Notes for Chapter 1

1. Jennifer Raff, "Finding the First Americans," *AEON Newsletter*, https://aeon.co/essays/the-first-americans-ahttps://aeon.co/essays/the-first-americans-a-story-of-wonderful-uncertain-science?utm_source=pocket-newtabstory-of-wonderful-uncertain-science?utm_source=pocket-newtab, accessed January 3, 2023.
2. Marc Thompson and Fred M. Morales, "El Paso: A Culture History and Urban Biography for the Union Plaza Redevelopment Program," chap. 2 in *The Union Plaza Downtown El Paso Development Archaeological Project: Overview, Inventory and Recommendations*, ed. Stephen Mbuto and John A. Peterson, ARC Archaeological Technical Report No. 17 (El Paso: Sun Metro Transit Authority, 1998), 29, 30.
3. Bradley J. Vierra, *Keystone in Context: A Significant Archaic Period Site in El Paso, Texas* (El Paso: Keystone Heritage Park, 2009), 1.
4. Rex W. Strickland, "Six Who Came to El Paso: Pioneers of the 1840's," *Southwestern Studies* 1, no. 3 (1963): 35.
5. Nancy González writes that "The Spanish government usually awarded land grants to soldiers who had served loyally, five consecutive terms of five years each term, in the Spanish Army." González, "Reinventing the Old West: Concordia Cemetery and the Power Over Space, 1800–1895" (PhD diss., University of Texas at El Paso, 2014), 28n55.
6. James Magoffin Dwyer Jr., "Hugh Stephenson," *New Mexico Historical Review* 29, no. 1 (1954): 4.
7. Jeffrey Marcos Garcilazo, *Traqueros: Mexican Railroad Workers in the United States, 1870 to 1930*, Al Filo: Mexican American Studies Series 6 (Denton: UNT Press, 2012), 11.
8. Oscar J. Martínez, *Stunted Dreams: How the United States Shaped Mexico's Destiny* (El Paso: El Paso Social Justice Education Project, 2017), 4.
9. Cleofas Calleros, "San José de Concordia el Alto," *El Paso World News*, June 4, 1932, 1–4.

10. W. H. Timmons, "American El Paso: The Formative Years," *Southwestern Historical Quarterly* 87, no. 1 (1983): 5.
11. Frank Louis Halla Jr., "El Paso, Texas, and Juárez, Mexico: A Study of Bi-Ethnic Community, 1846–1881" (PhD diss., University of Texas at Austin, 1978), 155.
12. Halla, "El Paso," 155.
13. Julia Lee Brown, "El Paso Cow Trails First Surveyed by Gen. Anson Mills," *El Paso Morning Times*, June 25, 1916, 1.
14. John Russell Bartlett, *Personal Narrative of Explorations and Incidents in Texas, New Mexico, California, Sonora, and Chihuahua: Connected with the United States and the United States Boundary Commission, During the Years 1850, '51, '52, and '53*, vol. 1 (London: George Routledge, 1854), 188–89.
15. Bartlett, *Personal Narrative*, 190–93.
16. According to W. H. Timmons, the California Column took possession of Fort Bliss after the Confederates had abandoned it on August 20, 1862. They remained in occupation until February 1865, when the United States Army took over. Timmons, *El Paso: A Borderlands History*, foreword by David. J. Weber (El Paso: University of Texas at El Paso, 1990), 178. Because Stephenson bought war bonds and was supportive of the Confederacy, Concordia was confiscated by the federal government, and his son-in-law Captain French purchased it at an estate sale and then divided it among his heirs.
17. Martin Donell Kohout, "Hugh Stephenson: Pioneer Settler and Trader in El Paso," *Handbook of Texas Online*, updated September 2, 2022, http://www.tshaonline.org/handbook/online/articles/fstcx.
18. Gonzalez, "Reinventing the Old West."
19. Dwyer, "Hugh Stephenson," 4.
20. Robert Neal Blake, "A History of the Catholic Church in El Paso" (MA thesis, College of the Mines, 1948), 35.
21. "Details for Site of Camp Concordia and Fort Bliss, Historical Marker – Atlas Number 5141004740," *Texas Historical Commission*, accessed July 20, 2025, https://atlas.thc.texas.gov/AdvancedSearch/HistoricalMarkerSearch.
22. Thomas Moore Carson, "Application to the Texas State Historical Commission for a Historical Marker for the Site of Camp Concordia and Fort Bliss Application," Marker Title: "Camp Concordia, Marker Number 4740, 4001 Durazno Street, El Paso, Texas," approved June 8, 1982, Texas Historical Commission, Austin.

23. Allan W. Sandstrum, "Fort Bliss: The Frontier Years" (MA thesis, University of Texas at El Paso, 1962), 120–22.
24. Sandstrum, "Fort Bliss," 122.
25. Sandstrum, "Fort Bliss," 124.
26. "Fort Bliss Exciting Century," *El Paso Herald-Post*, November 4, 1948, A1.
27. Welborn J. Williams Jr., "The Buffalo Soldiers' Brush with 'Jim Crow' in El Paso" (MA thesis, University of Texas at El Paso, 1996), 7.
28. Williams, "Buffalo Soldiers' Brush with 'Jim Crow,'" 8.
29. Williams, "Buffalo Soldiers' Brush with 'Jim Crow,'" 8.
30. Quintard Taylor, "Comrades of Color: Buffalo Soldiers in the West, 1866–1917," in *Western Voices: 125 Years of Colorado Writing*, ed. Steve Grinstead and Ben Fogelberg (New York: Fulcrum, 2004), 252–73.
31. Williams, "Buffalo Soldiers' Brush with 'Jim Crow,'" 2–3.
32. Williams, "Buffalo Soldiers' Brush with 'Jim Crow,'" 7.
33. Willis Newton, "Wagon Train—Texas to California 1865," accessed August 27, 2017, http://www.angelfire.com/ar/pyeatt/wagontrain.html.
34. Maceo Dailey Jr., "Border Black: The El Paso Story," in *Dígame! Policy & Politics on the Texas Border*, ed. Christine Thurlow Brenner, Irasema Coronado, and Dennis L. Soden (Dubuque, IA: Kendall/Hunt, 2003), 307–24.
35. Williams, "Buffalo Soldiers' Brush with "Jim Crow" in El Paso," 8.
36. Dailey, "Border Black," 310.
37. Dailey, "Border Black," 310.
38. Dailey, "Border Black," 308. According to Wellborn J. Williams, Black troops had served in El Paso since the end of the Civil War with two companies of the 125th United States Colored Troops, which were stationed at Fort Bliss; and between the Civil War and 1900, the 25th and 24th Infantry Regiments and the Ninth Calvary Regiment served in El Paso. Williams, "Buffalo Soldiers' Brush with 'Jim Crow,'" 7.
39. W. H. Timmons, "El Paso, Texas," in *The Portable Handbook of Texas*, ed. Roy R. Barkley and Mark F. Odintz (Austin: Texas State Historical Association, 2000), 309.
40. "Mr. And Mrs. Coffin Today Observe Their Golden Anniversary," *El Paso Times*, January 31, 1933.
41. "El Paso Earthquake of 1884 Vividly Recalled by Pioneer Woman," *El Paso Times*, October 9, 1937, p. 1, col. 6.
42. Allison Brownell Tirres, "Lawyers and the Legal Borderlands," *American Journal of Legal History* 50, no. 2 (2010): 158.

43. Joseph M. Cormack, "Legal Concepts in Cases of Eminent Domain," *Yale Law Journal* 41 (1931): 221.
44. Cormack, "Legal Concepts," 221–22.
45. Daniel B. Bendbow, "Public Use as a Limitation on the Power of Eminent Domain in Texas," *Texas Law Review* 44 (1965–1966): 1499.
46. Henry F. Triplett and Ferdinand A. Hauslein, *Civics: Texas and Federal* (Houston: Rein & Sons, 1912), 354.
47. Terry Cowan, "History of the Texas Public Domain" (Lecture, Texas Society of Professional Surveyors, Annual Convention & Technology Exposition, October 11, 2015), 16.
48. Cowan, "Texas Public Domain," 66.
49. Cowan, "Texas Public Domain," 67
50. Ricardo Romo, "The Urbanization of Southwestern Chicanos in the Early Twentieth Century," in *Chicano: The Evolution of a People*, ed. Renato Rosaldo, Robert A. Calvert, and Gustav L. Seligmann Jr. (Malabar, FL: Robert E. Krieger Publishing, 1982), 148.
51. Romo, "Urbanization of Southwestern Chicanos," 148.
52. Romo, "Urbanization of Southwestern Chicanos," 148.
53. Jo Ann Platt Hovious, "Social Change in Western Towns: El Paso, Texas, 1881–1889" (MA thesis, University of Texas, El Paso, 1972), 23–25.
54. Hovious, "Social Change," 13.
55. Mario T. García, *Desert Immigrants: The Mexicans of El Paso, 1880–1920*, Yale Western Americana Series 32 (New Haven: Yale University Press, 1981), 23.
56. Belinda Román, "Ciudad Juárez-El Paso: The Formation of a Cross-Border Market: Mexico/US Economic Relations in Perspective, 1840s–1920s" (PhD diss., London School of Economics and Political Science, 2003), 182–83, 242.
57. Rachel St. John, "Divided Ranges: Trans-Border Ranches and the Creation of National Space Along the Western Mexico-US Border," in *Bridging National Borders in North America*, ed. Benjamin Johnson and Andrew R. Graybill (Durham: Duke University Press, 2010), 118.
58. Digital Sanborn Maps, 1867–1970, Library of Congress, accessed October 22, 2017, https://www.loc.gov/collections/sanborn-maps/?ops=AND&qs=1908+fire+insurance+El+Paso&-searchType=advanced.
59. St. John, "Divided Ranges," 92.

60. El Paso Mission Trail Association, accessed October 22, 2017, https://www.facebook.com/EPMissionTrail/.
61. Román, "Ciudad Juárez-El Paso," 147.
62. Helen McCarthy Orndorff, "History of the Development of Agriculture in the El Paso Valley" (MA thesis, University of Texas at El Paso, 1957), 125.
63. Catherine Burnside O'Malley, "A History of El Paso Since 1860" (MA thesis, University of Southern California, 1939), 93. O'Malley stated, "Since 1902, El Paso has been a shipping point for cattle."
64. Orndorff, "Development of Agriculture," 125.
65. Gonzalez, "Reinventing the Old West."
66. El Paso County Official Public Records, accessed April 27, 2016, http://www.epcounty.com/records.htm.
67. Bureau of the Census, *Abstract of the Thirteenth Census of the United States Taken in the Year 1910 with Supplement for Texas* (Washington: Government Printing Office, 1913), 650.
68. Oscar J. Martínez, *Border Boom Town: Ciudad Juárez Since 1848* (Austin: University of Texas Press, 1975), 160.
69. García, *Desert Immigrants*, 44.
70. Figures vary, but according to historian Ricardo Romo US sources reported in 1916 that 17,198 Mexicans immigrated to the United States and Mexican sources reported the number at 49,932. "Table 1, Mexican Immigration to The United States 1910-1930" in Ricardo Romo, "Responses to Mexican Migration, 1910–1930," *Aztlán: A Journal of Chicano Studies* 6, no. 2 (Summer 1975): 178. It is important to note that not all Mexicans who migrated to the United States via El Paso stayed in the El Paso region.
71. Interview with Mrs. Oralee Smith by the author, El Paso, Texas, October 1, 2016.
72. "Census Report: El Paso, Texas, 1900–1990," Chase Bank of Texas in Celebration of Black History Month, February 2, 1998, El Paso, Texas, 27.
73. US Census Quick Facts, El Paso city, Texas, United States, accessed December 5, 2020, https://www.census.gov/quickfacts/fact/table/elpasocitytexas,US/RHI225219.
74. *El Paso City Directory, 1920* (Hudspeth Directory Co., 1920), 208, City Directories, Portal to Texas History, University of North Texas Libraries, Denton, accessed May 31, 2017, https://texashistory.unt.edu/ark:/67531/metapth285898/?q=El%20Paso%20City%20Directory%2C%201920.

Other Black churches in El Paso included Second Baptist Church, 401 S. Virginia (Rev. J. R. Jackson, pastor), and two Methodist Episcopal churches: Myrtle Avenue Church at 2023 Myrtle Ave. (Rev. C. Anderson, pastor) and Visitor's Chapel, 501 Tays Street (Rev. H. A. Wells, pastor).

75. "About/History," *NAACP El Paso Branch*, accessed February 22, 2018, http://naacpelpaso.org/history/.
76. Will Guzmán, "Appendix: List of NAACP Members," in "The El Paso Branch of the 1923 and 1929 National Association for the Advancement of Colored People," *Password: The El Paso County Historical Society* 60, no. 3 (Fall 2016): 85–87.
77. Ann R. Gabbert, "Defining the Boundaries of Care: Local Responses to Global Concerns in El Paso Public Health Policy, 1881–1941" (PhD diss., University of Texas at El Paso, 2006), 428. See also Shawn Lay, "Imperial Outpost on the Border: El Paso's Frontier Klan No. 100," in *The Invisible Empire in the West: Toward a New Historical Appraisal of the Ku Klux Klan of the 1920s*, ed. Shawn Lay (Urbana: University of Illinois Press, 1992), 67–96; and Lay, *War, Revolution, and the Ku Klux Klan* (El Paso: Texas Western Press, 1985).
78. Mary Bowling, "Lincoln Park School: A Brief History from Notes by Lillian Scott and 'The Clock,' a Fictional Memoir," in *Password: The El Paso County Historical Society* 39, no. 1 (Spring 1989): 25–26.
79. Bowling, "Lincoln Park School," 26–27.
80. Bowling, "Lincoln Park School," 27–28.
81. Letter to Hector González and the author from Max Grossman, PhD, vice chair, El Paso County Historical Commission, December 27, 2014.
82. El Paso Independent School District document, n.d., Open Record Request #2017.022, March 2, 2017, received March 20, 2017. In 1951 Lincoln School was remodeled with a four-room addition and an auditorium-lunchroom termed a Cafetorium.
83. Deed, the Board of Trustees of the Concordia Common School District No. 2, sale to the Independent School District of the City of El Paso, Texas, Volume 00416, page 00236-00237, El Paso County Records, accessed August 25, 2017, http://www.epcounty.com/publicrecords/officialpublicrecords/officialpublicrecordsearch.aspx.
84. Paul W. Horn, *Survey of the City Schools of El Paso, Texas* (El Paso: Department of Printing of the City Schools, 1922), accessed October 31, 2017, https://ia801404.us.archive.org/5/items/surveyofcityscho00hornrich/surveyofcityscho00hornrich.pdf.
85. Horn, *City Schools of El Paso*, 8–9.

86. El Paso Independent School District (EPISD) document, n.d., Open Record Request by the author, #2017.022, March 2, 2017, received March 20, 2017.
87. Bowling, "Lincoln Park School," 29. See also Miguel Juárez, "The Rich History of an El Paso Landmark," *Lincoln Park: El Paso's Chicano Park*, accessed July 14, 2015 https://lincolnparkcc.org/history/.
88. Bowling, "Lincoln Park School," 28.
89. Bowling, "Lincoln Park School," 32.
90. "Honor Roll in El Paso Schools (Continued from Page 2)," *El Paso Herald-Post*, Thursday, Dec. 5, 1935, 4.
91. The 1940 United States Federal Census (accessed December 10, 2016); available from Heritage Quest.
92. Manuel B. Ramírez, "When the Mexicans Go Home: El Paso and the Mexican Exodus, 1929–1934" (MA thesis, University of Texas at El Paso, 1994), 8.
93. A. O. Wynn, "A Study of the Operations of the El Paso Public Schools During the School Years 1930–31 Through 1945–46" (MA thesis, University of Texas at El Paso, 1946), 15.
94. According to an *El Paso Times* article titled "1955: Set Integration for Fall Term, First in Texas," June 22, 1955, El Paso Schools were desegregated in Fall 1955.
95. Gilbert G. González, *Chicano Education in the Era of Segregation* (Denton: University of North Texas Press, 1990), 2.
96. González, *Chicano Education*, 2–3.
97. González, *Chicano Education*, 46.
98. Blake, "Filial Church of Calvary," in "History of the Catholic Church," 98.
99. Church booklet, "*75 Años de Historia (Iglesia del Santo Angel), 50 Años de Servicio (Misioneras Claretianos)*," (Guardian Angel Church, 3021 Frutas, El Paso, Texas), 14–15, no date. See also Blake, "Filial Church of Calvary," in "History of the Catholic Church," 98.
100. Church booklet, "*75 Años,*" 14.
101. Church booklet, "*75 Años,*" 15.
102. Blake, "Filial Church of Calvary," in "History of the Catholic Church," 99.
103. Church booklet, "*75 Años.*"
104. Interview with David Prieto by the author, May 14, 2015, El Paso, Texas.
105. Miguel Juárez, Comments at the El Paso County Commissioners Meeting, March 9, 2021, regarding the Downtown 10 Project.

Notes for Chapter 2

1. Kathleen A. Tobin, "The Reduction of Urban Vulnerability: Revisiting 1950s American Suburbanization as Civil Defense," *Cold War History* 2, no. 2 (2002): 25.
2. Richard F. Weingroff, "Federal-Aid Highway Act of 1956: Creating the Interstate System," *Public Roads* 60, no. 1 (Summer 1996): 10. In a separate essay Weingroff writes, "In a presidential election year of 1956, a Congress controlled by Democrats combined with a Republican President to give the country a national highway network which President Dwight D. Eisenhower later spoke of as one of the greatest accomplishments of his 8 years in office. It is now named in his honor: The Dwight D. Eisenhower National System of Interstate and Defense Highways"; Weingroff, "A Vast System of Interconnected Highways: Before the Interstates," *Federal Highway Administration*, p. 1, accessed August 7, 2018, https://www.fhwa.dot.gov/highwayhistory/vast.pdf.
3. Tom Lewis argues, "It took forty years—not thirteen as specified by the legislation President Eisenhower signed in 1956—to build the Interstate System"; Lewis, *Divided Highways: Building the Interstate Highways, Transforming American Life* (Ithaca: Cornell University Press, 2013), xv.
4. Title VIII of the Civil Rights Act in 1968 prohibited discrimination in the sale, rental and financing of dwellings based on race, color, religion, sex, or national origin.
5. *Emerging Operations 1960, The Plan for El Paso, City Plan Commission*, p. 2, Jonathan R. Cunningham Papers, MS 287, Box 1, C. L. Sonnichsen Special Collections Department, UTEP. The State Highway Department and the City Traffic Engineering Department made a complete metropolitan area traffic survey. The Thorofare (*sic*) Plan evolved from the El Paso Highway Coordinating Committee and officially recognized by the State Highway Department as the basis for estimating and scheduling the state's construction program. The plan map was available to the public widely distributed.
6. Yolanda Chávez Leyva, "*Años de Desesparación*: The Great Depression and the Mexican American Generation in El Paso, Texas 1919–1935" (MA thesis, University of Texas at El Paso, 1989), 2.
7. Chávez Leyva, "*Años de Desesparación*," 3.
8. Chávez Leyva, "*Años de Desesparación*," 14.

9. Chávez Leyva, "*Años de Desesparación*," 4. The NRA was a governmental body created by Franklin D. Roosevelt to regulate employee rights. Deportation involved expelling someone who was seen as a foreigner based on illegal status or after having committed a crime, while repatriation is typically defined as the process of voluntarily returning to their place or country of origin.
10. Robert R. McKay, " Understanding Mexican Repatriation from Texas: Historical Insights," *Handbook of Texas Online*, updated April 8, 2020, http://www.tshaonline.org/handbook/online/articles/pqmyk.
11. Ramírez, "When the Mexicans Go Home," 8.
12. Ramírez, "When the Mexicans Go Home," 9.
13. Ramírez, "When the Mexicans Go Home," 15–16.
14. Manuel B. Ramírez, "El Pasoans: Life and Society in Mexican El Paso, 1920–1945" (PhD diss. University of Mississippi, December 2000), 148.
15. Gabbert, "Defining the Boundaries," 448.
16. Gabbert, "Defining the Boundaries," 8–9.
17. Gabbert, "Defining the Boundaries," 29.
18. Richard Rothstein, *The Color of Law: A Forgotten History of How Our Government Segregated America* (New York: Liveright, 2017).
19. "Rating of Location," Part II, 226-228, "Protection from Adverse Influences," Federal Housing Administration, Underwriting Manual (1936), excerpts. University of Michigan, (accessed date); available from https://epress.trincoll.edu/ontheline2015/wp-content/uploads/sites/16/2015/03/1936FHA-Underwriting.pdf.
20. Louis Lee Woods II, "The Federal Home Loan Bank Board, Redlining, and the National Proliferation of Racial Lending Discrimination, 1921–1950," *Journal of Urban History* 38 no. 6 (2012): 1036.
21. R. L. Olson, *Confidential Report of a Survey in El Paso, Texas for the Mortgage Rehabilitation Division, Home Owners' Loan Corporation*, Washington, DC: Home Owners' Loan Corporation, May 6, 1936.
22. El Paso HOLC Valuators included John O. Beckley, real estate, and management loan firm of Coles Brothers; J. Page Kemp, senior partner in the real estate firm of Kemp and Coldwell; R. F. Miller, vice president of Orndroff Realty Company, real estate brokerage and management firm; W. R. Piper, senior partner in the real estate, management, and loan firm of Marr-Piper Agency; J. E. Rogers, of the real estate and brokerage firm of Rogers & Belding; and Ray E. Sherman, mayor of El Paso and head of the real estate firm of Leavell & Sherman.

23. Olson, *Confidential Report*, "Summary of Interview with W. R. Piper," 2A.
24. Home Owners' Loan Corporation (HOLC) El Paso "Residential Security" map created from R. M. Metcalfe Co. Map Publishers, "Map and Street Guide of El Paso, Texas," El Paso, Tex. Edition, No. 779, 1936, Record Group 195: Records of the Federal Home Loan Bank Board, 1933–1989, Box 154, Folder for El Paso, Texas, National Archives and Records Administration, College Park, Maryland.
25. Olson, "Confidential Report," 11.
26. Olson, "Confidential Report," 11. It was as if someone took a marker and colored the HOLC report map by hand or laid colored acetates on it.
27. HOLC map, 1936.
28. Olson, "Confidential Report," Description of Areas, 2.
29. County of El Paso, Official Public Records, https://apps.epcounty.com/publicrecords/OfficialPublicRecords, accessed April 2021, and City of El Paso Planning Department, Alex Hoffman, Chief City Planner.
30. City of El Paso Planning Department, Alex Hoffman, Chief City Planner.
31. County of El Paso County Public Records, https://www.epcounty.com/records.htm, accessed May 10, 2021.
32. Raymond A. Mohl, "Urban Expressways and the Central Cities in Postwar America," in *The Interstates and the Cities: Highways, Housing, and the Freeway Revolt* (Poverty and Race Research and Action Council, 2002), 2.
33. Avila, *Popular Culture*, 34.

Notes for Chapter 3

1. Interview with Jacqueline D. Hoyt by the author, February 7, 2025.
2. Interview with Lydia and Mateo Hinojosa by the author, April 20, 2015.
3. Hinojosa interview.
4. Hinojosa interview.
5. The El Paso Builders Association had offices in Bassett Towers and was created by George C. Hervey (brother of former El Paso city mayor Fred Hervey), who served as their first president. The association also published a magazine. An inquiry by the author into the ElPaso Builder's Association's archives produced no materials. Current officers are unaware of any materials or archives from the association's early years.

6. Hinojosa interview.
7. Hinojosa interview.
8. Hinojosa interview.
9. Becky Sterling Interview by the author, September 8, 2010. Yards were removed from the residents to build the gateways for Interstate 10. In building highways states had a choice on building gateways. In Texas, highway builders created gateways (also called freeway access roads).
10. "Former Pupils Honor Teacher," *El Paso Herald-Post*, Friday, October 23, 1959, 20.
11. Mauro Rosas, c.v., n.d., courtesy of his nephews and brothers Alex and Roberto "Robi" Rosas.
12. Zachary Foust and R. Matt Abigail, "Mauro Rosas: Pioneering Mexican American Attorney and State Representative," *Handbook of Texas Online*, September 27, 2016, http://www.tshaonline.org/handbook/online/articles/froef.
13. Interview with Alex and Robert "Robi" Rosas by the author in El Paso, Texas, on June 27, 2016.
14. Rosas interview.
15. John P. Schmal, "The Tejano Struggle for Representation," *The Hispanic Experience: Hispanics in Government*, Houston Institute for Culture, Special Feature, accessed November 7, 2011, http://www.houstonculture.org/hispanic/tejanorepprint.html.
16. Rosas interview.
17. Rosas interview.
18. Department of Commerce—Bureau of the Census, Sixteenth Census of the United States: 1940 Population Schedule, J. Pct. #1, Block Nos. 523, Lorraine C. Cole, Enumerator (accessed March 25, 2017); available from Heritage Quest.
19. Interview with H. L. Scales by the author, El Paso, Texas, September 27, 2016.
20. Interview with Mrs. Oralee Smith by the author, El Paso, Texas, October 1, 2016. The Buffalo soldier Mrs. Smith referred to might have been Donnie Wah Brown, who was born on April 8, 1908, and passed away June 1, 1995. He lived at 3203 Wyoming Street in the East El Paso Addition east of the Lincoln Park neighborhood. He was a master sargent and private in the US Army (discharged in 1953). He served in World War II and in Korea. He is buried at the Section O Site 1878 at the Fort Bliss National Cemetery. Mrs. Smith stated during

her interview that Mr. Brown, as she called him (not knowing his first name), is buried at the Fort Bliss National Cemetery, and he is one of nine Buffalo soldiers buried there.

21. Scales interview.
22. El Paso County Deed Records, accessed October 10, 2016, epcounty.com.
23. Smith interview. Mrs. Smith stated a live-in job was one where an individual worked Monday through Friday at a residence and on weekends they went home and returned on Monday.
24. Smith interview.
25. Smith interview.
26. Smith interview.
27. Ray Eli Rojas, "Under I-10: Life in Lincoln and La Roca," *Newspaper Tree*, January 3, 2005.
28. Rojas, "Under I-10."
29. Smith interview.
30. Interview with David Prieto by the author, May 14, 2015.
31. Interview with Francisco Prieto, by the author, March 24, 2015, El Paso, Texas.
32. D. Prieto interview.
33. D. Prieto interview.
34. D. Prieto interview.
35. Interview with Dalia Prieto-Rivero by the author, April 21, 2015.
36. "Lincoln School Must Be Moved for Highway," *El Paso Herald-Post*, July 17, 1965, A1, A2, Col. 6. According to Chihuahuita resident Freddy Morales, a protest occurred against highway building in 1978 as depicted in a mural by Chihuahuita residents titled "Untitled" or "A Tribute to Joe Battle" or "Lágrimas" on the 200 block of Montestruc, formerly Seventh Avenue. The mural was painted in 1978 by Fred Morales and Mara (full name unknown) and other Chihuahuita residents. It addressed the possibility of TxDOT running a freeway through the neighborhood, which it later did. The mural was sponsored by Proyecto Bello (Beautification Project) under Project Bravo Southside. Sandra Ivette Enriquez, "¡El *Barrio Unido Jamás Será Vencido!* Neighborhood Grassroots Activism and Community Preservation in El Paso, Texas" (PhD diss., University of Houston, 2016).
37. County of El Paso, Official Public Records, accessed April 2021, https://apps.epcounty.com/publicrecords/OfficialPublicRecords.
38. United States Census Bureau, accessed July 26, 2024, www.census.gov.

39. Telephone Interview with Pedro Martínez by the author, July 3, 2017.
40. El Paso Black History Tour (African American Historic Sites), published by the El Paso Community Foundation, 2025.
41. Facebook e-mail chat with Mari Primero, former Lincoln Park resident, by the author, December 20, 2016.
42. A talented artist in high school who created the banners for the football team, Salvador Leal Jr. was drafted to Vietnam right after high school. He was killed by friendly fire.
43. According to an interview of Richard Telles by José Angel Gutíerrez in *Tejano Voices*, June 22, 1996, http://library.uta.edu/tejanovoices/interview.php?cmasno=034: "Telles was born in 1922, and was raised in El Paso and served as El Paso City Clerk, Democratic Party Precinct Chairman, President of the American National Insurance Council, and El Paso County Commissioner. He was a member of the Masons, the Political Association of Spanish-Speaking Organizations (PASSO), the League of United Latin American Citizens (LULAC) and helped to form Viva Kennedy Clubs in El Paso." At the time of his death in 2003, Richard Telles was president of the El Paso Independent School District board of trustees, the first Mexican American in the role in the last century.
44. Oscar J. Martínez, *The Chicanos of El Paso: A Case of Changing Colonization*, Southwestern Studies 59 (El Paso: Texas Western Press, 1980), 11. A series of articles published in the *El Paso Times* in December 1968 found similar absences of Mexican Americans as lawyers, as presidents and committee members of the El Paso Chamber of Commerce and the Sun Carnival Parade, and as homeowners in affluent neighborhoods. The City of El Paso Planning Department Papers, MS 204, Box 23, Folder 11. C. L. Sonnichsen Special Collections Department, UTEP.
45. Interview with Rosa Guerrero by the author, September 26, 2016, El Paso, Texas.

Notes for Chapter 4

1. Scott Campbell, Urban Planning 540, University of Michigan, "Planning History Timeline: A Selected Chronology of Events (with a Focus on the US), accessed August 4, 2016, http://wwwpersonal.umich.edu/%7Esdcamp/up540/timeline12.html#prog.
2. Raymond A. Mohl, "Urban Expressways," 1.

3. Raymond A. Mohl, "Planned Destruction: The Interstates and Central City Housing," in Raymond A. Mohl and Roger Biles, eds., *The Making of Urban America*, 3rd ed. (Lanham: Rowman & Littlefield Publishing), 289.
4. Herbert J. Gans, "The Failure of Urban Renewal," in *American Urban History: An Interpretive Reader with Commentaries*, ed. Alexander B. Callow Jr. (New York: Oxford University Press, 1982), 455.
5. Grigsby, "Housing and Slum Clearance," 107–18.
6. Schermbeck, *Urban Renewal for Texas*, 3.
7. Jon C. Teaford, "Urban Renewal and Its Aftermath," *Housing Policy Debate* 11, no. 2 (2000): 443, http://dx.doi.org/10.1080/10511482.2000.9521373.
8. Teaford, "Urban Renewal and Its Aftermath," 445.
9. Teaford, "Urban Renewal and Its Aftermath," 445.
10. Judson F. Williams, "President's Message: Urban Affairs Are Local Affairs," *El Paso Today*, November 6, 1961, p. 1.
11. "Giant Step Interstate System," *Texas Highways*, January 1964, 10–11.
12. The National System of Interstate and Defense Highways Improvement Status of System Mileage as of September 1963 in "Interstate: More Than the Sum of Its Parts," *Texas Highways*, February 1964, 13.
13. "The White House, Statement by the President of the Formation of a Committee to Rebuild America's Slums," Office of the White House Press Secretary, June 2, 1967, Executive, LBJ Presidential Library and Archives, FG 647, Folder: Committee to Rebuild America's Slums (11/22/63—12/31/67), [1 of 3].
14. "Statement by the President."
15. Colin Gordon, "Blighting the Way: Urban Renewal, Economic Development, and the Elusive Definition of Blight," *Fordham Urban Law Journal* 31, no. 2 (2003): 316.
16. Tobin, "Reduction of Urban Vulnerability," 4.
17. *The Historical Overview of Housing in El Paso*, 46th Annual Report (El Paso City Plan Commission, 1972), 6.
18. *Historical Overview*, 7.
19. *Historical Overview*, 8.
20. *Historical Overview*, 8.
21. *Historical Overview*, 6.
22. David R. Díaz. *Barrio Urbanism: Chicanos, Planning and American Cities* (New York: Routledge, 2005), 163.
23. Peter Schrag, *El Paso Times*, May 12, 1954, p. A1.

24. The El Paso Annexation Map approved by Charles W. Davis, P. E., city engineer, City of El Paso Engineering Department. Map drawn by S. F. Jr., October 1965.
25. Hervey Construction built hundreds of low-priced homes between Hawkins to the west, North Loop to the south, Lee Trevino to the east, and Interstate 10 to the north, as well as in east and northeast El Paso. George C. Hervey was one of the founders of the El Paso Builder's Association. An October 6, 1967, article in the *El Paso Herald-Post* stated Hervey had been active in homebuilding and land development in east and northeast El Paso and had built more than 2,500 homes during his career as an El Paso builder; "Meet Parade of Homes Builders," *El Paso Herald-Post*, Friday, October 6, 1967, Section D, 3.
26. Lydia Otero, *La Calle: Spatial Conflicts and Urban Renewal in a Southwest City* (Tucson: University of Arizona Press, 2010), 97.
27. Sarah F. Liebschutz, "Neighborhood Revitalization in the United States: The Decentralization Dynamic," *Public Administration Quarterly* 14, no. 1 (Spring 1990): 88.
28. Schrag, *El Paso Times*, p. A1.
29. The Housing Act of 1949, 42 USC. 1941 (as amended).
30. Robert E. Lang and Rebecca R. Sohmer, "Legacy of the Housing Act of 1949: The Past, Present, and Future of Federal Housing and Urban Policy," *Housing Policy Debate* 11, no. 2 (2000): 291, http://dx.doi.org/10.1080/10511482.2000.9521369.
31. Interview with Mateo Hinojosa by the author, El Paso, Texas, April 20, 2015.
32. Jewel Bellush and Murray Hausknecht, eds., *Urban Renewal: People, Politics, and Planning* (Garden City, NY: Anchor Books, Doubleday, 1967), 275.
33. MACHOS stood for Mexican American Committee on Honor Opportunity and in Project BRAVO, BRAVO stood for Building Resources and Vocational Opportunities.
34. *Workable Program for Community Improvement* (Washington, DC: Housing and Home Finance Agency, July 1962), 2.
35. *Workable Program*, 3.
36. Project Bravo, CAP-81, (Alameda Policy Advisory Committee, 1965), MS 2014, Box 52, Folder #14, p. 13, City of El Paso Planning Department Papers, MS 204, Box 25, Folder 19, C. L. Sonnichsen Special Collections Department, UTEP.

37. Schermbeck, *Urban Renewal for Texas*, 39.
38. "People and Land Use Patterns," in *El Paso Development Manual, Urban Development Manual for the City of El Paso: A Handbook for Community Planning*, prepared by Robert E. Alexander, (F.A.I.A. & Associates, Architects and Planning Consultants, Los Angeles, CA, May 1968), 49, ACC#880 El Paso Document Collection, Box 13, Folder *El Paso Development Manual, Urban Development Manual for the City of El Paso, A Handbook for Community Planning,* C. L. Sonnichsen Special Collections Department, UTEP.
39. City of El Paso Community Development Program, 1975, (Department of Planning and Research, City of El Paso, Texas, March 1975), n.p. "First Year funding was $100,000, Second Year $150,000 and Third Year $50,000," City of El Paso Planning Department Papers, MS 204, Box 24, Folder 5, C. L. Sonnichsen Special Collections Department, UTEP.
40. Articles of Incorporation of El Paso Community Action Program Project BRAVO, approved April 14, 1965, Article Four, page 1, Upward Bound and Project BRAVO, Dr. Ralph Segalman, MS 001, Box 1, Unprocessed Collection, C. L. Sonnichsen Special Collections Department, UTEP.
41. Ralph Segalman, "An Application for Funds Under Public Law 88-452, The Economic Opportunity Act of 1964, Project Bravo (Building Resources and Vocational Opportunities) Community Action Program, El Paso County, El Paso, Texas, March 1965, p. 6, MS 001, Box 1 of 1, No Folder, C. L. Sonnichsen Special Collections Department, UTEP.
42. Segalman, "Application for Funds," 6.
43. Segalman, "Application for Funds," 6.
44. Segalman, "Application for Funds," 28.
45. Segalman, "Application for Funds," 28.
46. Segalman, "Application for Funds," 11.
47. Segalman, "Application for Funds," 11.
48. Segalman, "Application for Funds," 11.

Notes for Chapter 5

1. Leon C. Metz, *Turning Points in El Paso, Texas* (El Paso: Mangan Books, 1985), 108.
2. Interview with Attorney Tom Diamond in his residence by the author, August 5, 2016.

3. Jeffrey M. Schulze, “The Chamizal Blues: El Paso, the Wayward River, and the Peoples in Between,” *Western Historical Quarterly* 43 (Autumn 2012): 303
4. Schulze, “Chamizal Blues,” 310.
5. Letter from F. C. Turner, Director of Public Roads, US Department of Transportation, Federal Highway Administration, Bureau of Public Roads, Washington, DC, 20591, to J. C. Dingwall, State Highway Engineer, Texas Highway Department, Austin, Texas 78703, February 29, 1968, EX HI (Executive Highway), Box 1, p. 1, Papers of Lyndon Baines Johnson, President, 1963–1969, LBJ Presidential Library, Austin (hereafter cited as LBJ Papers).
6. Correspondence to Mr. and Mrs. Leopold Patino by the Texas Highway Department, District 24, regarding the creation of the Chamizal Border Freeway, November 13, 1970, El Paso County, Control 2552-4, CBH L375 (2), Parcel 148, Michael Patino Family Papers.
7. Correspondence to Mr. and Mrs. Leopold Patino by the Texas Highway Department, October 21, 1970, Patino Family Papers.
8. Interview with Tom Diamond by the author, August 5, 2016.
9. Maria Eugenia Trillo, “The Code-Switching Patterns of Rio Linda Community of El Chamizal in El Paso, Texas: An Emic Perspective of Syntactic Constraints” (PhD diss., University of New Mexico, 2002), 13.
10. Trillo, “Code-Switching Patterns,” 13.
11. Trillo, “Code-Switching Patterns,” 13.
12. Trillo, “Code-Switching Patterns,” 13.
13. “Highway Committee Aids Multi-Million Dollar Program,” *El Paso Today,* April 1967, 4.
14. *El Paso Today*, November 6, 1961, 1.
15. An overtly positive editorial titled “Ending the Chamizal Question,” on the Chamizal dispute, appeared in 1962 in *El Paso Today*, August 1, 1962, p. 2, but four years later, when residents challenged property appraisals in court, the issue was viewed as a problem: “Chamizal Project Construction Near,” *El Paso Today*, March 1966, p. 8.
16. “El Pasoans Support Chamizal Settlement Plan,” *El Paso Today,* August 1, 1962, 3.
17. Leon C. Metz, “Restless River, Disputed Land,” in *Turning Points*, 113.
18. Arturo and Vallarie Enriquez, *Vila* (Vantage Point Visual Studios, 2006). I would like to thank the Enriquezes for loaning me their video on Vila.
19. Arturo and Vallarie Enriquez, *Vila.*

20. Marshall Hail, "'Little People' Due to Have Hearing on Chamizal Plan," *El Paso Herald-Post*, Wednesday, February 20, 1963, 1A-6A, Col. 1.
21. For further information on Elvira Villa Lacarra Escajeda see: Alana de Hinojosa, "Elvira Villa Lacarra Escajeda: Pioneer of Community Advocacy in El Paso," *Handbook of Texas Online*, updated June 20, 2022, https://www.tshaonline.org/handbook/entries/escajeda-elvira-villa-lacarra.
22. *La Compra Del Derecho De Via*, Departamento de Carreteras de Texas (sp), Impreso por el Departamento de Planificación de la Ciudad de El Paso en Cooperación con El Departamento de Carreteras de Texas (sp), n.d., City of El Paso Planning Department Papers, MS 204, Box 6, Folder #20, *"La Compra Del Derecho* (Right-of- Way) de Via," C. L. Sonnichsen Special Collections Department, UTEP.
23. *La Compra Del Derecho*, 5.
24. *La Compra Del Derecho,* 13.
25. *La Compra Del Derecho*, 19.
26. Personal conversation with Historian David D. Romo, July 15, 2016, El Paso, Texas.
27. Both President Dwight D. Eisenhower and Thomas Harris MacDonald, chief of the Federal Bureau of Public Roads in the mid-1920s, had been members of AASHO. In his 2013 book *Divided Highways*, Tom Lewis writes that MacDonald was present at the group's formation in 1914.
28. John David Huddleston, "Good Roads for Texas: A History of the Texas Highway Department, 1917–1947" (PhD diss., Texas A&M University, 1981), iii.
29. Robert E. Ireland, "Prison Reform, Road Building, and Southern Progressivism: Joseph Hyde Pratt and the Campaign for 'Good Roads and Good Men,'" *North Carolina Historical Review* 68, no. 2 (April 1991): 125, http://www.jstor.org/stable/23521190.
30. Mohl, "Planned Destruction," 289.
31. Mohl, "Planned Destruction," 290. Mohl states Thomas H. MacDonald, a highway engineer from Iowa, headed the Bureau of Public Roads from its founding until 1953 and he relentlessly promoted his agency's road-building agenda.
32. *General Location of National System of Interstate Highways, including all Additional Routes at Urban Areas Designated in September 1955*, US Department of Commerce, Bureau of Public Roads, accessed August 15, 2016, https://archive.org/stream/generallocationo00unitrich#page/n0/mode/2up.

33. Lewis, *Divided Highways*, 120.
34. According to the Federal Highway Administration Highway History site, "The 'Yellow Book,' [was] the influential 1955 publication by the US Bureau of Public Roads that contains maps showing the general routing of urban Interstates . . . the formal title was *General Location of National System of Interstate Highways Including All Additional Routes at Urban Areas Designated in September 1955*," accessed February 26, 2017, https://www.fhwa.dot.gov/interstate/links.cfm.
35. Tobin, "Reduction of Urban Vulnerability," 4.
36. Charles B. Quattlebaum, "Military Highways," *Military Affairs* 8 (January 1944): 225, 230.
37. Tobin, "Reduction of Urban Vulnerability," 25.
38. *Texas Highway Commission Public Hearing on Regional Development*, El Paso, Texas, December 1, 1966, Exhibit Number 7.00, 1 of 1, 16. Highways were crucial for moving military equipment in the 1900s, but the surfaces could not handle the loads and roads deteriorated.
39. "El Paso to Host National Defense Seminar," *El Paso Today,* July 1, 1961, pp. 1, 4.
40. "Record-Smashing Security Seminar Opens: Top Attendance for Security Course Expected," *El Paso Today*, November 1, 1961, p. 1.
41. Eric Avila, "Betty, Barbara, Joan, and Jane: The Gendered Dimensions of Highway Construction in Postwar America," University of Minnesota, Quadrant Lecture Series, funded by the Andrew W. Mellon Foundation, April 28, 2009, http://quadrant.umn.edu/post/projects/2010/folklore-freeway-cultural-history-highway-construction/presentation-betty-barbara.
42. Huddleston, "Good Roads for Texas," iii.
43. William A. Bugge, "The 1956 Federal Aid Highway Act from the State Viewpoint," *American Highways*, April 1957, p. 6.
44. D. C. Greer, State Highway Engineer, Letter: Administrative Order No. 23-56 to Division Heads, District Engineers, and Engineer-Managers, Subject: Organization and Recruiting, September 26, 1956, Box 2002/101-8, Highway Commission Letters, 1954–1970, Texas State Archives, Lorenzo de Zavala State Archives and Building Library, Capitol Complex, Austin, Texas. On September 19, 1956, the Texas Highway Commission passed Minute Order No. 40217, approving a comprehensive Right-of-Way Program allocating funds for the securing of right-of-way on specified sections of the Interstate System.

45. The Austin office had their in-house magazine called *Highway*. Issues of *Highway* contained employee news and articles, baby photographs, promotions, retirements, and deaths.
46. Hosmer W. Hill, El Paso, Texas, Letter to D. C. Greer, State Highway Engineer, February 5, 1954, Box 2002/101-8, Commission Letters, 1954–1970, Texas State Archives.
47. "A Brief Geological History of the El Paso-Juárez, ELP-J Region," accessed September 26, 2017, http://www.geo.utep.edu/loca/fieldtrip.html
48. Hill to Greer, February 5, 1954, Commission Letters, 1954–1970, Texas State Archives.
49. Hill to Greer, February 5, 1954.
50. State of Texas, County of El Paso, Deed of Exchange between Hosmer W. Hill and Anna G. Hill and El Paso County, Texas, July 10, 1957 (accessed July 2, 2017); available from epcounty.com.
51. Year: *1940; Census Place: El Paso, El Paso, Texas;* Roll: *T627_4183;* Page: *13A;* Enumeration District: *256-87* Ancestry.com. *1940 United States Federal Census* [database on-line]. Provo, UT, USA, (accessed March 3, 2018); available from Ancestry.com.
52. "El Paso Promised Fund for Freeway, Highway Board Willing to Spend $50 Million Here, Projects Advanced at Meeting With El Paso Officials," *El Paso Herald-Post*, September 20, 1955, A3.
53. "Mayor Favors Report on Total Cost of Freeway," *El Paso Herald-Post*, September 22, 1955, A1.
54. "Mayor Favors Report."
55. "Hearing Set on Freeway Links," *El Paso Herald-Post*, Saturday, June 22, 1957, A1.
56. Bonnie Resler Karlsrud, *Dale Resler Behind the Scenes in El Paso: A Historical Narrative* (pub. by author, 2007), 124.
57. "Membership Committee Named by Williams," *El Paso Today,* April 17, 1961, p. 1. A May 17, 1961, issue of the magazine listed transportation as an increasingly important aspect in the Southwest for the Chamber committee.
58. Martínez, *Chicanos of El Paso*, 11.
59. Interview with retired attorney Tom Diamond by the author, August 5, 2016.
60. The White House, Washington, Memorandum for the President, Subject: Revised Estimate of Cost of Interstate Highway System, from Lee C. White, January 12, 1965, Lyndon Baines

Johnson Library & Museum, Box 1, (EX HI) Executive Highway, 11/22/61, Papers of LBJ.

61. "State Set to Spend $32 Million In El Paso," *El Paso Herald-Post*, October 2, 1964, Section A, 1.
62. Mohl, *Interstates and the Cities*, 3.

Notes for Chapter 6

1. "On Highway Land Costs: Appraiser Asks State to Figure Full Value," *Our Southwest Section, El Paso Herald-Post*, December 25, 1957, 12.
2. Frank L. White, "Downtown Interests Pushing Plans for Depressed Freeway," *El Paso Herald-Post*, April 23, 1959, A1.
3. "Downtown Bypass Road Considered by State," *El Paso Herald-Post*, Thursday, January 24, 1957, 3.
4. "Freeway Route Now Defined," *El Paso Herald-Post*, Tuesday, May 6, 1958, A1.
5. Miguel Juárez, with photographs by Cynthia Weber Farah, *Colors on Desert Walls: The Murals of El Paso* (El Paso: Texas Western Press, 1997), 11.
6. "Loop 375 Border West Extension Project," *Texas Department of Transportation*, accessed April 19, 2021, https://www.txdot.gov/insidehttps://www.txdot.gov/inside-txdot/projects/studies/el-paso/border-highway- west.htmltxdot/projects/studies/el-paso/border-highway-west.html.
7. Ellwyn R. Stoddard, *The Role of Social Factors in the Successful Adjustment of Mexican American Families to Forced Housing Relocation: A Final Report of the Chamizal Relocation Research Project, El Paso, Texas* (Department of Planning Research and Development, Community Renewal Program, City of El Paso, 1970). El Paso Community Renewal Program, *Relocation Practices in El Paso*, 1971. The preparation of this report financed in part through a Community Renewal Program Grant from the Department of Housing and Urban Development as authorized by Title I of the Housing Act of 1949, as amended, Project No. TEX R-132 (CR), Department of Planning and Research Community Renewal Program, City of El Paso, 1971, Vertical files, Urban Renewal Folder, City of El Paso Planning Department Papers, El Paso Public Library.
8. Stoddard, *Role of Social Factors*, iv.

9. Community Renewal Program, *Relocation Practices in El Paso.*
10. Raymond A. Mohl, "Stop the Road: Freeway Revolts in American Cities," *Journal of Urban History* 30, no. 5 (July 2002): 674–706, https://doi.org/10.1177/0096144204265180 (Original work published 2004. An article titled "Chamizal Project Construction Near" in *El Paso Today*, March 1966, p. 8, reported, "The US International Boundary and Water Commission has released a Chamizal Project Progress Report concerning the convention for settlement of the Problem of the Chamizal. Land Acquisition: 626 properties were acquired or 755 of the total number of vacant units. Most of the properties acquired to date are residential. Purchase contracts signed amount to $6.18 million. Approved were 99 of the 110 claims filed by residents for special compensation, amounting to $54,864. They were referred to the Justice Department 27 cases for price determination. The program contemplates essential completion of the land acquisition by or soon after July 1966."
11. Interview with William E. Wood Jr. by Michelle L. Gomilla, Institute for Oral History, the University of Texas at El Paso, 17. .
12. Interview with Wood by Gomilla, 17.
13. "Ask Mrs. Carroll, 'Can Anyone Remember E.P. When It Was Sleepy Town?'" *El Paso Herald-Post*, Monday, February 8, 1965, Section C, p. 2
14. Mindy Thompson Fullilove, *Root Shock: How Tearing Up City Neighborhoods Hurts America, and What We Can Do About It* (New York: New Village Press, 2016), 4.
15. Ramírez, "El Pasoans," 135–36.
16. Barbara Lee, *Renegade for Peace and Justice: Congresswoman Barbara Lee Speaks for Me* (Lanham, MD: Rowman & Littlefield, 2008), 9.
17. Interview with Alex and Robert "Robi" Rosas by the author in El Paso, Texas, on June 27, 2016.
18. David Montejano, "The Demise of 'Jim Crow' for Texas Mexicans, 1940–1970," *Aztlán: A Journal of Chicano Studies* 16, nos. 1–2 (1985): 30. See also David Montejano, *Anglos and Mexicans in the Making of Texas, 1836–1986* (Austin: University of Texas Press, 1987), 31.
19. Montejano, *Anglos and Mexicans*, 32.
20. "Multiple Use Right of Way Agreement," "Right-of-Way for the Multiple Land Use Park in the IH-10 and IH-110 Interchange," State of Texas, County of El Paso, August 8, 1974, 1. In an email message from Alan Shubert, City Engineer, to Deborah G. Hamlyn,

Qualify of Life Director, City of El Paso, Subject: 4001 Durazno Lincoln Center Summary, August 2, 2011, in a three-ring binder titled "Lincoln Center," prepared by the LPCC, Shubert states that the city sent a letter to the state requesting the use of state-owned right-of-way beneath Interstate 110 structures for a public purpose: to create Lincoln Park.

21. Correspondence from Bruce Yetter, Assistant City Attorney, Legal Department, the City of El Paso, to Jonathan R. Cunningham, Director of Planning and Research, City of El Paso, Re: Lincoln School Renovation Project, May 14, 1975, Jonathan R. Cunningham Papers, MS 287, C. L. Sonnichsen Special Collections Department, UTEP.
22. E-mail from Shubert to Hamlyn, August 2, 2011.
23. Wayne McClintock, "City Authorizes Federal Contract," *El Paso Herald-Post*, June 23, 1975, Section A2.
24. City of El Paso Presentation at City Council on Lincoln Center, October 4, 2011, in a three-ring binder titled "Lincoln Center," prepared by LPCC. According to Jon C. Teaford in his article "Urban Renewal and Its Aftermath," CDBG, created by Congress in 1974, took over where urban renewal funds left off. Teaford states that "the CDBG scheme sought to give localities greater flexibility in using federal funds.
25. Community Development Program 1975 prepared and designed by the staff of the Department of Planning and Research of the Public Information Requirement of the Community Development Act, printed by the El Paso City Printing Department, (Department of Planning and Research, City of El Paso, Texas, March 1975), 4.
26. Frank G. Kelly, "Plans Are Ridiculous," *Texas Highways*, June 1970, 8–9.
27. Kelly, "Plans Are Ridiculous," 8.
28. Marisol Rodríguez and Hector Rivero, "ProNaF, Ciudad Juárez: Planning and Urban Transformation," *ITU Journal of the Faculty of Architecture* 8, no. 1 (2011): 196.
29. Amy Sara Carroll, *Remex: Toward an Art History of the NAFTA Era* (Austin: University of Texas Press, 2017), 210.
30. Carroll, *Remex*, 210.
31. Carroll, *Remex*, 199.
32. Carroll, *Remex*, 210.
33. Carroll, *Remex*, 204.
34. Carroll, *Remex*, 198.
35. Oscar J. Martínez, *Troublesome Border*, rev. ed. (Tucson: University of Arizona Press, 2006), 111.

36. Erika Esquivel, "Rep. Escobar Announces $600 Million for Expansion of Bridge of the Americas Port of Entry," *KFox14*, March 25, 2022, https://kfoxtv.com/news/immigration/rep-escobar-announces-600-million-for-expansion-of-bridge-of-the-americas-port-of-entry.
37. National Historic Preservation Act of 1966 (as amended through 1992) Public Law 102-575, Title I, Section 106 (16 USC. 470f), accessed August 10, 2018, https://www.nps.gov/history/local-law/nhpa1966.htm.
38. Maryellen Russo, "Historical Resources Survey Report, Reconnaissance and Intensive Survey Report," I-10 Connect, Yandell Drive to Loop 375 (Cesar Chavez Border Highway), El Paso District, El Paso County, June 2018.
39. The National Historic Preservation Act, which is a federal law, mandates that "for projects using federal funds, licenses, or permits, project planners mist consult with potentially interested parties, including the State Historic Preservation Office (SHPO) to: (1) identify historic properties potentially affected by an undertaking; and (2) and avoid, minimize, or mitigate adverse effects on historic properties." It is not known whether such a study was conducted on El Calvario Church prior to it being torn down in 1970. Texas Capital Fund Workshop, Introduction to Section 106 Review, Section 106 Overview: The Regulations, accessed April 26, 2018, http://www.thc.texas.gov/public/upload/Texas%20Capital%20FundTexasHistoricalCommissionTraining201.
40. A black-and-white aerial photograph dated 1961 shows the construction of Bassett Center as early as 1961 while the roads are being cleared for the creation of Interstate 10.
41. Correspondence from Joe M. Battle, Texas Highway Department, District Engineer, to the Honorable Fred Hervey, Mayor, City of El Paso, Subject: Multiple Land Use Agreement in IH-10 and US 54 Interchange, September 10, 1974, Texas Highway Department Historical Records, Texas State Archives.
42. E-mail message from Alan Shubert, City Engineer, to Deborah G. Hamlyn, Qualify of Life Director, City of El Paso, Subject: 4001 Durazno Lincoln Center Summary, August 2, 2011.
43. "Highway Building Booms in El Paso," *Texas Highways*, November 1966, 5.
44. Texas Highway Commission Meeting Minutes, January 29, 1964–January 29, 1970, Lorenzo de Zavala State Archives and Building Library, Texas State Archives, Capitol Complex, Austin, Texas.

45. "Aids Multi-Million Dollar Program," *El Paso Today*, April 1967, 4.
46. For a discussion of highways as military efforts, see Charles B. Quattlebaum, "Military Highways," 225; Kathleen A. Tobin, "Reduction of Urban Vulnerability."

Notes for Chapter 7

1. Vic Kolenc, "Medical Center of the Americas' Plan Seeks Downtown-like development in Central El Paso," *El Paso Times*, October 4, 2018, https://www.elpasotimes.com/story/news/health/2018/10/04/medical-center-americas-plan-aims-building-second-el-paso-downtown/1385008002/.
2. El Paso City Manager Joyce Wilson's presentation on "History of Lincoln School," October 18, 2011, p. 5, City Council Meeting Minutes, City of El Paso.
3. The Juntos Art Association was founded as the National Association of Chicano Arts (NACA) in late 1985. Members later changed its name to the Juntos Art Association by the extraordinarily successful Juntos Art Exhibits.
4. Lilia Patricia Reade de Pellicano, "*Abuelitas de El Paso*: A Selective Cultural Study" (MA thesis, University of Texas at El Paso, 1989), 33–36.
5. Interview with Hector González and Gabriel Gaytán by the author, August 13, 2016.
6. González and Gaytán interview.
7. Timothy W. Collins, "Marginalization, Facilitation, and the Production of Unequal Risk: The 2006 *Paso del Norte* Floods," Antipode 42, no. 2 (March 2010): 279, doi: 10.1111/j.1467- 8330.2009.00755.x.
8. Wilson presentation on "History of Lincoln School," October 18, 2011, p. 7, City Council Meeting Minutes, City of El Paso.
9. González and Gaytán interview, August 13, 2016.
10. Correspondence from John Garza, City of El Paso Environmental Services, to Louis Aloysius, *Microbial Investigation on Lincoln Center*, October 16, 2006, Lincoln Center, City of El Paso.
11. Garza, *Microbial Investigation*, 6.
12. Wilson's presentation on "History of Lincoln School," October 18, 2011, p. 7, City Council Meeting Minutes, City of El Paso.
13. "Comments about Lincoln Center," in e-mail correspondence from Joyce A. Wilson, City Manager, to Nanette L. Smejkal, Director of El

Paso Parks and Recreation, January 02, 2008, Planning Department, City of El Paso.

14. E-correspondence regarding "Lincoln Ctr. via Freedom of Information Request FOIA," p. 1.
15. "Lincoln Ctr. via Freedom of Information Request FOIA," 1. Joyce Wilson was the first city manager hired when the City of El Paso went into a city manager form of governance under then Mayor Joe Wardy. Wilson had been city manager for six to seven months when Lincoln Center was closed. At that point Wilson was under the impression that the city owned Lincoln Center, which was incorrect because TxDOT had owned it and had leased it to the city.
16. "Lincoln Ctr. via Freedom of Information Request FOIA," 1.
17. Dubbed as the "Storm 2006," on July 27, 2006, heavy rains caused widespread flood in El Paso County. The city received fifteen inches of rain, which caused widespread damage to homes and businesses. "10 Years Ago Storm 2006 Damaged Hundreds of Homes, Caused Hundreds of Millions in Damage," KVIA, News 7, August 01, 2016, http://www.kvia.com/news/10-years-ago-storm-2006-damaged-hundreds-of-homes-caused-hundreds-of-millions-in-damage/89182130.
18. Timthy W. Collins, "The Production of Unequal Risk in Hazardscapes: An Explanatory Frame Applied to Disaster at the US-Mexico Border," *Geoforum* 40 (2009): 590.
19. González and Gaytán interview, August 13, 2016.
20. In August 2006 heavy rains occurred in El Paso, Texas. During the rains Lincoln Center was used as an emergency shelter for residents of the Saipan-Ledo neighborhood, where fifty families were evacuated. After the flood in October 2006, citing mold, Lincoln Center was closed by then City Manager Joyce Wilson. Wilson stated that the 2006 flood was the reason for the mold although it was due to the city's negligence and not the flood.
21. Wilson's presentation on Lincoln Center Proposed Demolition, October 5, 2011.
22. City of El Paso, Historic Preservation Officer (Providencia Velázquez) Analysis of Lincoln Center, n.d.
23. Analysis of Lincoln Center.
24. Lincoln Center FOIA cover letter by Raymundo Eli Rojas, June 2, 2011.
25. Three-ring binder titled "Lincoln Center," prepared by LPCC for El Paso City Representative Emma Acosta, 2015.

26. Initially, representing the district where Lincoln Park is located, El Paso City Representative Emma Acosta was supportive against the demolition of Lincoln Center, but over time she was no longer in support of the project. In addition, due to redistricting, the Lincoln Park neighborhood was no longer part of her authority, so she did not feel it was within her purview to support the reopening of the center.
27. Correspondence from Hector González, President, Lincoln Park Conservation Committee, to Texas Senator José Rodríguez, May 10, 2011, Texas Highway Department Historical Records, 2011–2015, Texas State Archives.
28. "Hector Gutierrez, Jr., '69," Corps of Cadets, Texas A&M University, accessed September 7, 2016, http://corps.tamu.edu/portfolio-items/hector-gutierrez/.
29. Dr. Mariana Chew, Lincoln Center Assessment, Xicali Engineering, El Paso, Texas, June 12, 2013.
30. El Paso Community College, *Site Evaluation of 4001 Durazno–Lincoln Center by the EPCC Physical Plant, Police Department, and Information Technology Department, 8-12-2013*, El Paso, Texas, 2.
31. Texas Department of Transportation, Multiple Use Agreement, Form 2044 (Rev. 07/2012), 3, TxDOT Archives.
32. Multiple Use Agreement, 4.
33. Multiple Use Agreement, 5–6.
34. Lincoln Park Conservation Committee sent a letter to TxDOT Chairperson Houghton on August 13, 2013, with electronic copies to Senator Rodriguez, state and city representatives, and Dr. Dennis Bixler Marquez, Director of Chicano Studies at UTEP.
35. "Real Estate Management and Development," Texas Department of Transportation, accessed April 29, 2018, https://www.txdot.gov/business/opportunities/real-estate.html.
36. Marissa Luck, "Prime Austin Land Near Busy Intersection Offered to Developers for Nearly $8 million," *Austin Business Journal*, March 20, 2018, https://www.bizjournals.com/austin/news/2018/03/20/prime-austin-land-near-busy-intersection-offered.html.
37. David Crowder, "Stop Ahead: Rebuilding I-10," *El Paso Inc.*, January 26, 2014.
38. Miguel Juárez and Corinne Chacón, *Business Plan: Lincoln Center for Chicano Cultural Arts, Wellness and Community Archive*, Lincoln Park Conservation Committee and The Senecu Fine Arts Society, Prepared

for Mayor Oscar Leeser and El Paso City Council Representatives, April 23, 2014.

39. Texas Transportation of Transportation Meeting, June 26, 2014, Pasadena, TX, http://txdot.swagit.com/play/06262014-688.
40. Texas Transportation of Transportation Meeting, October 30, 2014, El Paso, TX http://txdot.swagit.com/play/10302014-548.
41. "Saving Lincoln Center: An El Paso Community's Effort to Protect Its Heritage," *National Trust for Historic Preservation*, accessed July 25, 2014, http://blog.preservationnation.org/2014/07/25/saving-lincolnhttp://blog.preservationnation.org/2014/07/25/saving-lincoln-center-el-paso-communitys-effort-protect- heritage/center-el-paso-communitys-effort-protect-heritage/#.U-xo2-d1bLQ.
42. Artist Celia Muñoz, e-mail correspondence, February 2014.
43. Interview with Carlos Callejo, by the author, September 7, 2008.
44. Hal Marcus, "El Paso's Lincoln Center Will Be Saved: 10 Reasons by Hal M," *YouTube*, May 26, 2014, https://www.youtube.com/watch?v=r-uA1HlOVdM.
45. Miguel Juárez, in Draft Application for the United States Department of the Interior, National Park Service, National Register of Historic Places Registration, Criteria G: A property achieving significance within the past 50 years if it is of exceptional importance, 15-16.
46. Organized in 2009, the Lincoln Park Conservation Committee, or LPCC, is a city-recognized neighborhood association for the Lincoln Park community. LPCC has also been designated by the City of El Paso as a Partner in Parks and is the only grassroots organization at the table with TxDOT in the state.
47. Texas Department of Transportation, Maryellen Russo, Principal Investigator, *Historical Resources Survey Report, Reconnaissance and Intensive Survey: I-10 Connect Project, Yandell Drive to Loop 375 (Cesar Chavez Border Highway), El Paso District, El Paso County*, June 2018.
48. Aaron Montes, "El Paso TxDOT District Engineer Bob Bielek Resigns Abruptly," *El Paso Times*, May 15, 2019, https://www.elpasotimes.com/story/news/2019/05/15/el-paso-txdot-district-engineer-bobbielek-resigns/3684236002/.
49. "Trevino Named New District Engineer for El Paso TxDOT," *El Paso Herald Post*, July 10, 2019, accessed July 22, 2019, https://elpasoheraldpost.com/trevino-named-new-district-engineer-for- el-paso-txdot/.

Notes for Conclusion

1. "The Trench," is a TxDOT term used by its engineers for covering the freeway at downtown El Paso and building new entrance and exit ramps through downtown and adjacent neighborhoods in proximity through downtown El Paso.
2. Aaron Bracamontes, "Former El Paso City Manager Joyce Wilson Appointed to Board by Texas Gov. Greg Abbott," KTSM.com, April 23, 2019, https://www.ktsm.com/local/el-paso-news/former-el-paso-city-manager-joyce-wilson-appointed-to-board-by-texas-gov-greg-abbott.
3. Vic Kolenc, "Bridge Parks, Gateways, and Demolition. How Would the I-10 Expansion Affect Downtown El Paso?" *El Paso Times*, May 30, 2019, https://www.elpasotimes.com/story/news/2019/05/30/i-10expansion-concept-would-add-bridge-parks- demolish-downtown-el-paso-buildings/3686340002/
4. "Fact Sheet: The Bipartisan Infrastructure Deal," The White House, November 06, 2021, https://bidenwhitehouse.archives.gov/briefing-room/statements-releases/2021/11/06/fact-sheet-the-bipartisan-infrastructure-deal/.
5. "Secretary Buttigeig: Infrastructure Updates Must Be Equitable," March 18, 2021, Amanpur & Co., https://www.pbs.org/wnet/amanpour-and-company/video/secy-buttigeig-infrastructure-updates-must-be-equitable-ard0/.
6. The author was selected to be a consulting party to the Section 106 process. There was one online meeting.
7. Dr. Kathy Staudt, Facebook post in El Paso Street Coalition.

Notes for the Appendix

1. María Cortés, "City Names Loop 375 After Battle," *El Paso Times*, October 9, 1991, in Trish Long, "Tales from the Morgue," April 23, 2008, http://elpasotimes.typepad.com/morgue/2008/04/joe-battle.html.
2. In 1961 construction proceeded on a $244 million master highway program with the opening of several major interchanges in Houston, Texas. Houston History Timeline, accessed May 10, 2023, https://02db39d.netsolhost.com/decades/timeline/5m1tl.htm,.
3. *Texas Highways*, September 1963, 3.
4. *Texas Highways*, September 1963, 3.
5. Cortés, "City Names Loop 375 After Battle."

6. Interview with Attorney Tom Diamond in his residence by the author, August 5, 2016. Diamond stated Bill Reiger was a right-of-way purchasing agent for Interstate 10 in 1958. Formerly from Austin, Reiger relocated to El Paso to join the right-of-way office. Efforts to interview Mr. Reiger, the last surviving right-of-way agent for the construction of Interstate 10, were futile.
7. "Job Still a Joy to El Paso's Battle: A Conversation with El Paso District Engineer Joe Battle," *Transportation News: State Department of Highways and Public Transportation*, February 1990, p. 4.
8. Diamond interview, p. 2.
9. Diamond interview, p. 2.
10. Diamond interview, p. 3.
11. Diamond interview, p. 3.
12. Diamond interview, p. 3.
13. Interview with Joan Cunningham-Estrada by the author, El Paso, Texas, March 20, 2018.
14. Cunningham-Estrada interview.
15. Cunningham-Estrada interview.
16. Biography from a finding guide to the Jonathan R. Cunningham Papers, MS 287, C. L. Sonnichsen Special Collections Department, UTEP.
17. Cunningham Papers at the C. L. Sonnichsen Special Collections Department is limited to fifty inches, or ten small boxes, and does not reflect the twenty-year career of a planner during the height of freeway building in El Paso, Texas. Another larger collection, the El Paso Planning Department Papers, MS 204, also at the C. L. Sonnichsen Special Collections Department, does include about 100 boxes of materials, but the collection has many elements, as well as some duplications, and items such as deeds and contracts, are not included. Former Director of El Paso City Planning Nestor Valencia, in a private conversation, stated the City of El Paso maintains a city archive with thousands of boxes in East El Paso that are not open to the public, except under Freedom of Information (FOIA) request. There is no public finding aid for this collection.
18. Cunningham arrived in El Paso after serving as director of planning for Spokane County in Washington from 1952 to 1956. He studied public administration and planning at the University of Chicago. As a professional planner, he was committed to contributing to the profession and often authored articles about planning that he presented at planning conferences.

19. The White House, Chamizal Memorial Grounds, "Invitation to Mr. and Mrs. Jon Cunningham, El Paso, Texas, October 29, 1977."
20. "First Ladies Bridge Cultures," *El Paso Herald Post*, November 3, 1977, A1.
21. "First Ladies Bridge Cultures,"
22. Cunningham-Estrada interview.
23. Resume of Nestor A. Valencia, former director of El Paso City Planning, 2007, 1.
24. Díaz states that in 1965, the first major book on a Latina/o urbanism was Patricia Cayo- Sexton's *Spanish Harlem*, and it "was an early model of how planning addressed Latinas/os and spatial relations." Yet after its publication, he states, "the planning literature focused almost exclusively on Afro-Americans, and this minority group served as the explanatory model for all minority urbanisms." Díaz, *Barrio Urbanism*, 5.
25. Cunningham-Estrada interview.
26. Interview with Manuel F. Aguilera by the author.
27. Bill Lockhart, "Bottles on the Border: The History and Bottles of the Soft Drink Industry in El Paso, Texas, 1881–2000," Revised Edition, 2010, accessed October 8, 2017, https://sha.org/bottle/pdffiles/EPChap7a.pdf. According to Lockhart, the Woodlawn Addition was named for the Woodlawn Bottling Company. The bottling company opened on Frutas Avenue in 1909 under the proprietorship of Martin R. Sweeney and John J. McArdle (240).
28. Aguilera interview.
29. Aguilera verified Lincoln School had been a field office for the Highway Department to build I-10.
30. Aguilera interview.
31. Aguilera interview.
32. Aguilera interview.
33. Aguilera interview. In December 2018 the El Paso portion of Interstate 10 turned fifty years old.
34. Aguilera interview
35. A traffic engineer makes sure that there is traffic flow, that signs are set up during construction, to keep commuters informed and that people get to where they want to go. Aguilera interview.
36. "Highway Renaming Plan Draws Emotional Protest from a Namesake, Family," KVIA, August 24, 2015, http://www.kvia.com/news/highway-renaming-plan-draws-emotional-protest-from- namesake-family/56195559.

Bibliography

Archives

C. L. Sonnichsen Special Collections Department, University of Texas at El Paso Library.

El Paso Department of Planning Collection, 1923–1999

Jonathan R. Cunningham Papers, 1925–1983

City Directories. Portal to Texas History, University of North Texas Libraries, Denton.

City of El Paso, Texas.

City Council Archived Meeting Minutes, Municipal Clerk's Office. http://www.elpasotexas.gov/muni_clerk/archived_minutes.asp.

Lincoln Center

Planning Department

Court of Civil Appeals of Texas, El Paso

El Paso Independent School District, El Paso

El Paso Public Library

Border Heritage Center

City of El Paso Planning Department Papers

Urban Renewal Folder

Institute for Oral History. University of Texas at El Paso.

Lincoln Park Conservation Committee (LPCC) Papers, 2005–2016, El Paso, Texas.

Lyndon B. Johnson Presidential Library and Archives, Austin, Texas.

Papers of Lyndon Baines Johnson President, 1963–1969

Michael Patino Personal Family Papers.

National Archives and Records Administration, College Park, Maryland.

Texas Department of Transportation Archives, Austin, Texas.

Texas Historical Commission, Austin.

Texas State Archives, Austin.

Texas Highway Commission Letters, 1954–1970

Texas Highway Commission Meeting Minutes, January 29, 1964–January 29, 1970

Texas Highway Commission Minutes, 1963–1975

Texas Highway Department Historical Records, 1911–1993

Texas Highway Department Historical Records, 2011–2015

Newspapers and Magazines

American Highways
El Paso Herald-Post
El Paso Inc.
El Paso Morning Times
El Paso Times
El Paso Today
El Paso World News
Newspaper Tree
Orion Afield
Texas Highways
Transportation News

Other Sources

Altshuler, Alan, and David Luberoff. *Mega-Projects: The Changing Politics of Urban Public Investment*. Washington, DC: Brookings Institution Press and Lincoln Institute of Land Policy, 2003.

Anderson, Martin. *The Federal Bulldozer: A Critical Analysis of Urban Renewal*. Cambridge: MIT Press, 1964.

Archer, John. "Social Theory of Space: Architecture and the Production of Self, Culture, and Society." *Journal of the Society of Architectural Historian* 64, no. 4 (December 2005): 430–33.

Arreola, Daniel D., and James R. Curtis. *The Mexican Border Cities: Landscape Anatomy and Place Personality*. Tucson: University of Arizona Press, 1993.

Avila, Eric. "Betty, Barbara, Joan, and Jane: The Gendered Dimensions of Highway Construction in Postwar America." University of Minnesota, Quadrant Lecture Series, funded by the Andrew W. Mellon Foundation, April 28, 2009.

Avila, Eric. "The Folklore of the Freeway: Space, Identity and Culture in Postwar Los Angeles." *Aztlán: A Journal of Chicano Studies* 23, no. 1 (Spring 1998): 13–31.

Avila, Eric. *Popular Culture in the Age of White Flight: Fear and Fantasy in Suburban Los Angeles*. Berkeley: University of California Press, 2004.

Avila, Eric. "Popular Culture in the Age of White Flight: Film Noir, Disneyland, and the Cold War (Sub)Urban Imaginary." *Journal of Urban History* 31, no. 1 (2004): 3–22.

Avila, Eric. “Revisiting the Chavez Ravine: Baseball, Urban Renewal and the Gendered Civic Culture of Postwar Los Angeles.” In de Alba, *Velvet Barrios*.

Avila, Eric, and Mark Rose. “Race, Culture, Politics and Urban Renewal: An Introduction.” In “Race, Culture, Politics and Urban Renewal,” ed. Eric Avila and Mark Rose, special issue, *Journal of Urban History* 35, no. 3 (March 2009): 335–47.

Avila, H. J. “Immigration and Integration: The Mexican American Community in Garden City, Kansas 1900–1950.” *Kansas History* 20 (1997): 22–37.

Baldwin, Davarian L. *Chicago's New Negroes: Modernity, the Great Migration, and Black Urban Life.* Durham: University of North Carolina Press, 2007.

Barber, Llana. *Latino City: Immigration and Urban Crisis in Lawrence, Massachusetts, 1945–2000*. Durham: University of North Carolina Press, 2017.

Barr, Alwyn. *Black Texans: A History of African Americans in Texas, 1528–1995*. Norman: University of Oklahoma Press, 1996.

Bartlett, John Russell. *Personal Narrative of Explorations and Incidents in Texas, New Mexico, California, Sonora, and Chihuahua: Connected with the United States and the United States Boundary Commission, During the Years 1850, '51, '52, and '53*. Vol. 1. London: George Routledge, 1854.

Bauman, Robert. “The Black Power and Chicano Movements in the Poverty Wars in Los Angeles.” *Journal of Urban History* 33 (January 2007): 277–95.

Behnken, Brian D. *The Struggle in Black and Brown: African American and Mexican American Relations During the Civil Rights Era.* Lincoln: University of Nebraska Press, 2011.

Bellush, Jewell, and Murray Hausknecht, eds. *Urban Renewal: People, Politics, and Planning.* Garden City, NY: Anchor Books, Doubleday, 1967.

Bendbow, Daniel B. “Public Use as a Limitation on the Power of Eminent Domain in Texas.” *Texas Law Review* 44 (1965–1966): 1499.

Bhabha, Homi K. *The Location of Culture*. New York: Routledge, 1994.

Blake, Robert Neal. “A History of the Catholic Church in El Paso.” MA thesis, College of the Mines, 1948.

Boehm, Lisa Krissoff. “Adding Gender to American Urban History.” *Journal of Urban History* 36 (January 2010): 56–67.

Bowden, J. J. “The Spanish Ascarate Grant.” MA thesis, Texas Western College, 1952.

Bowden, J. J. *Spanish and Mexican Land Grants in the Chihuahuan Acquisition*. El Paso: Texas Western Press, 1971.

Bowling, Mary. "Lincoln Park School: A Brief History from Notes by Lillian Scott, and 'The Clock,' a Fictional Memoir." *Password: The El Paso County Historical Society* 39, no. 1 (Spring 1989): 25–36.

Brandt, Stefan L. "The City as Liminal Space: Urban Visuality and Aesthetic Experience in Postmodern US Literature and Cinema." *Amerifastudien / American Studies* 54, no. 4 (2009): 553–81.

Brown, Norman D. *Hood, Bonnet, and Little Brown Jug: Texas Politics, 1921–1928*. College Station: Texas A&M University Press, 1984.

Bullard, Robert D. "Addressing Urban Transportation Equity in the United States." *Fordham Urban Law Journal* 31, no. 5 (October 2004): 1183–210.

Bullard, Robert D. "All Transit Is Not Created Equal." *Race, Poverty, and the Environment* 12, no. 1 (Winter 2005/2006): 9–12.

Bullard, Robert D. "Confronting Environmental Racism in the 21st Century." In *The Colors of Nature: Culture, Identity, and the Natural World*, edited by Alison H. Deming and Lauret E. Savoy. Minneapolis: Milkweed Editions, 2002.

Bullard, Robert D. "Just Transportation in the Era of Suburban Sprawl." *Journal of Community Development* 39, no. 3 (2008).

Bullard, Robert D., and Glenn S. Johnson. "Environmental Justice: Grassroots Activism and Its Impact on Public Policy Decision Making." *Journal of Social Issues* 56 (2000): 555–78.

Bullard, Robert D., and Glenn S. Johnson, eds. *Just Transportation: Dismantling Race and Class Barriers to Mobility.* Stony Creek, CT: New Society, 1997.

Bullard, Robert D., Glenn S. Johnson, and A. O. Torres. "Atlanta: Megasprawl." *Forum for Applied Research and Public Policy* (Fall 1999): 17–23.

Bullard, Robert D., Glenn S. Johnson, Angel O. Torres. "Building Transportation Equity into Smart Growth." In *Highway Robbery: Transportation Racism & New Routes to Equity*, edited by Bullard, Johnson, and Torres. Cambridge, MA: South End Press, 2004.

Bullard, Robert D., G. S. Johnson, and A. O. Torres. "The Costs and Consequences of Suburban Sprawl: The Case of Metro Atlanta." *Georgia State University Law Review* 17 (Summer 2001): 935–98.

Bullard, Robert D., Glenn S. Johnson, and Angel O. Torres. "Dismantling Transportation Apartheid." *American Bar Association Human Rights Magazine* 34, no. 3 (Summer 2007): 2–6.

Bullard, Robert D., G. S. Johnson, and A. O. Torres. "The Routes of Transportation Apartheid." *Forum for Applied Research and Public Policy* 15 (Fall 2000): 66–4.

Bureau of the Census. *Abstract of the Thirteenth Census of the United States Taken in the Year 1910 with Supplement for Texas*. Washington, DC: Government Printing Office, 1913.

Callow, Alexander B., Jr., ed. *American Urban History: An Interpretative Reader with Commentaries*. New York: Oxford University Press, 1982.

Camarillo, Albert M. "Chicano Urban History: A Study of Compton's Barrio, 1936–1970." *Aztlan: A Journal of Chicano Studies* 2, no. 2 (1971): 79–106.

Camarillo, Albert M. *Chicanos in a Changing Society: From Mexican Pueblos to American Barrios in Santa Barbara and Southern California, 1848–1930*. Dallas: Southern Methodist University Press, 2005.

Camarillo, Albert M. "Navigating Segregated Life in America's Racial Borderhoods, 1910–1950s." *Journal of American History* 100 (December 2013): 645–63.

Campa, Arthur. "Immigrant Latinos and Resident Mexican Americans in Garden City, Kansas: Ethnicity and Ethnic Relations." *Urban Anthropology and Studies of Cultural Systems and World Economic Development* 19, no. 4 (Winter 1990): 345–60.

Carroll, Amy Sara. *Remex: Toward an Art History of the NAFTA Era*. Austin: University of Texas Press, 2017.

Chávez Leyva, Yolanda. "*Años de Desesparación*: The Great Depression and the Mexican American Generation in El Paso, Texas 1919–1935." MA thesis, University of Texas at El Paso, 1989.

Chernoff, Michael Light. "The Social Impacts of Urban Sections of the Interstate Highway System." PhD diss., University of Massachusetts, 1976.

Christian, Garna Loy. "Sword and Plowshare: The Symbiotic Development of Fort Bliss and El Paso, Texas, 1849–1918." PhD diss., Texas Tech University, 2017.

Cockcroft, Eva. "The Story of Chicano Park." *Aztlán: A Journal of Chicano Studies* 15, no. 1 (September 1984): 79–103.

Collins, Timothy W. "Marginalization, Facilitation, and the Production of Unequal Risk: The 2006 *Paso del Norte* Floods." *Antipode* 42, no. 2 (March 2010): 258–88.

Collins, Timothy W. "The Production of Unequal Risk in Hazardscapes: An Explanatory Frame Applied to Disaster at the US-Mexico Border." *Geoforum* 40 (2009): 589–601.

Contreras, Eduardo. "Voice and Property: Latinos, White Conservatives, and Urban Renewal in 1960s San Francisco." *Western Historical Quarterly* 45, no. 3 (Autumn 2014): 253–76.

"Convict Labor for Road Work." *Monthly Review of the U.S. Bureau of Labor Statistics* 4, no. 4 (1917): 591–95.

Cormack, Joseph M. "Legal Concepts in Cases of Eminent Domain." *Yale Law Journal* 41 (1931): 221.

A Correct Map of the United States Showing the Union Pacific, the Overland Route, and Connections. Map. West Texas excerpt. Chicago: Knight, Leonard, 1892.

Cowan, Terry, "History of the Texas Public Domain." Lecture, Texas Society of Professional Surveyors, Annual Convention & Technology Exposition, October 11, 2015.

Dailey, Maceo, Jr. "Border Black: The El Paso Story." In *Dígame! Policy & Politics on the Texas Border*, edited by Christine Thurlow Brenner, Irasema Coronado, and Dennis L. Soden. Dubuque, IA: Kendall/Hunt, 2003.

Davis, J. F. "United States Population Changes, 1960–1970." *Geography* 57, no. 2 (April 1972): 140–44.

Davis, Mike. *City of Quartz: Excavating the Future in Los Angeles*. New York: Vintage, 1992.

Dawson, Ronald E. "Streetcars at the Pass, Vol. 1: The Story of the Mule Cars of El Paso, the Suburban Railway to Tobin Place, and the Interurban to Ysleta." *Journal of the Railroad & Transportation Museum of El Paso*, no. 1 (2003): 77.

de Alba, Alicia Gaspar, ed. *Velvet Barrios: Popular Culture and Chicana/o Sexualities*. New York: Palgrave Macmillan, 2003.

De La Garza, Rudolph O. "Voting Patterns in 'Bi-Cultural' El Paso: A Contextual Analysis of Chicano Voting Behavior." *Aztlán: A Journal of Chicano Studies* 5, nos. 1–2 (Spring–Fall 1974): 235–60.

Del Castillo, Richard Griswold, ed. *Chicano San Diego: Cultural Space and the Struggle for Justice.* Tucson: University of Arizona Press, 2007.

Del Castillo, Richard Griswold. *The Treaty of Guadalupe Hidalgo: A Legacy of Conflict.* Norman: University of Oklahoma Press, 1990.

Díaz, David R. *Barrio Urbanism: Chicanos, Planning and American Cities*. New York: Routledge, 2005.

Dwyer, James Magoffin, Jr. "Hugh Stephenson." *New Mexico Historical Review* 29, no. 1 (1954): 4.

Ebner, Michael H. "Re-Reading Suburban America: Urban Population De-concentration, 1810–1980." *American Quarterly* 37, no. 3 (1985): 368–81.

Edensor, Tim, and Mark Jayne. *Urban Theory Beyond the West: A World of Cities*. New York: Routledge: 2012.

Enriquez, Arturo, and Vallarie Enriquez. *Vila*. Vantage Point Visual Studios, 2006.

Enriquez, Sandra Ivette. "¡El *Barrio Unido Jamás Será Vencido!* Neighborhood Grassroots Activism and Community Preservation in El Paso, Texas." PhD diss., University of Houston, 2016.

Epstein, Richard. *Takings: Private Property and the Power of Eminent Domain*. Cambridge: Harvard University Press, 1985.

Fairbanks, Robert B. "The Failure of Urban Renewal in the Southwest: From City Needs to Individual Rights." *Western Historical Quarterly* 37, no. 3 (Autumn 2006): 303–25.

Fairbanks, Robert B. *War on Slums in the Southwest: Public Housing and Slum Clearance in Texas, Arizona, and New Mexico, 1935–1965*. Philadelphia: Temple University Press, 2014.

Feagin, Joe R. *Urban Revitalization and Displacement: Types, Causes, and Public Policy*. Austin: Center for Energy Studies, University of Texas at Austin, 1981.

Foust, Zachary, and R. Matt Abigail. "Mauro Rosas: Pioneering Mexican American Attorney and State Representative." *Handbook of Texas Online*, September 27, 2016. http://www.tshaonline.org/handbook/online/articles/froef.

Friedkin, J. F., United States Commissioner. *A Preliminary Report on United States Problems in Acquisition of Private Properties in Connection with the Proposed Chamizal Settlement*. International Boundary and Water Commission, August 2, 1963.

Fullilove, Mindy Thompson. *Root Shock: How Tearing Up City Neighborhoods Hurts America, and What We Can Do About It*. New York: New Village Press, 2016.

Gabbert, Ann R. "Defining the Boundaries of Care: Local Responses to Global Concerns in El Paso Public Health Policy, 1881–1941." PhD diss., University of Texas at El Paso, 2006.

Gans, Herbert J. "The Failure of Urban Renewal." In Callow, *American Urban History*.

García, Mario T. *Desert Immigrants: The Mexicans of El Paso, 1880–1920*. Yale Western Americana Series 32. New Haven: Yale University Press, 1981.

Garcilazo, Jeffrey Marcos. *Traqueros: Mexican Railroad Workers in the United States, 1870 to 1930*. Al Filo: Mexican American Studies Series 6. Denton: UNT Press, 2012.

Garfinkle, Ann. *The Legal and Ethical Consideration of Mural Conservation: Issues and Debates*. Los Angeles: Getty Conservation Institute, 2003.

Gibson, Campbell, and Kay Jung. *Historical Census Statistics of the Foreign-Born Population of the United States: 1850–2000, Working Paper No. 81*. Washington, DC: US Bureau of the Census, 2000.

Gonzalez, Erualdo R. *Latino City: Urban Planning, Politics, and the Grassroots*. New York: Routledge, 2017.

González, Gilbert G. *Chicano Education in the Era of Segregation*. Denton: University of North Texas Press, 1990.

Gonzalez, Nancy. "Reinventing the Old West: Concordia Cemetery and the Power Over Space, 1800–1895." PhD diss., University of Texas at El Paso, 2014.

Gordon, Colin. "Blighting the Way: Urban Renewal, Economic Development, and the Elusive Definition of Blight." *Fordham Urban Law Journal* 31, no. 2 (2003): 316.

Gordon, Colin. *Mapping Decline: St. Louis and the Fate of the American City*. Philadelphia: University of Pennsylvania Press, 2008.

Gottfreund, Owen D. *20th-Century Sprawl: Highways and the Reshaping of the American Landscape*. Oxford: Oxford University Press, 2004.

Grigsby, William G. "Housing and Slum Clearance: Elusive Goals." *Annals of the American Academy of Political and Social Science* 352 (March 1964): 107–18.

Gruen, Claude. "Urban Renewal's Role in the Genesis of Tomorrow's Slums." *Land Economics* 39, no. 3 (August 1963): 285–91.

Gunning, Michael N. "An Annotated Bibliography of Urban Planning and Related Materials Available at the University of Texas at El Paso Library." MA thesis, University of Texas at El Paso, 1973.

Guzmán, Will. *Civil Rights in the Texas Borderlands: Dr. Lawrence A. Nixon and Black Activism*. Urbana: University of Illinois Press, 2015.

Guzmán, Will. "The El Paso Branch of the 1923 and 1929 National Association for the Advancement of Colored People." *Password: The El Paso County Historical Society* 60, no. 3 (Fall 2016): 85–87

Halla, Frank Louis, Jr. "El Paso, Texas, and Juárez, Mexico: A Study of Bi-Ethnic Community, 1846–1881." PhD diss., University of Texas at Austin, 1978.

Harris, Charles H., III, and Louis R. Sadler. *Bastion on the Border: Fort Bliss, 1854–1943*. Historical and Natural Resources Report no. 6. Fort Bliss, TX: Cultural Resources Management Branch, Directorate of Environment, US Army Air Defense Artillery Center, 1993.

Hayden, Dolores. *The Power of Place: Urban Landscapes as Public History*. Cambridge: MIT Press, 1995.

Hayden, Dolores. *Redesigning the American Dream: Gender, Housing and Family Life*. New York: W. W. Norton, 2002.

Henig, Jeffrey R. *Neighborhood Mobilization, Redevelopment, and Response*. New Brunswick: Rutgers University Press, 1982.

Henry, Jim. "Preservation Plan for Lincoln Center, College of Architecture and Planning." MA thesis, University of Arizona–Tucson, 2015.

Hester, Randolph. *Neighborhood Space*. Stroudsburg, PA: Dowden, Hutchinson, and Ross, 1975.

de Hinojosa, Alana. "Elvira Villa Lacarra Escajeda: Pioneer of Community Advocacy in El Paso." *Handbook of Texas Online*, updated June 20, 2022. https://www.tshaonline.org/handbook/entries/escajeda-elvira-villa-lacarra.

de Hinojosa, Alana. "El Rio Grande as Pedagogy: The Unruly, Unresolved Terrains of the Chamizal Land Dispute." *American Quarterly* 73, no. 4 (2021).

The Historical Overview of Housing in El Paso. 46th Annual Report. El Paso: El Paso City Plan Commission, 1972.

Hodges, Gladys Arlene. "El Paso, Texas and Ciudad Juárez, Chihuahua, 1880–1930: A Materials Culture Study of Borderlands Interdependency." PhD diss., University of Texas at El Paso, 2010.

Hooks, Greg, and Chad L. Smith. "The Treadmill of Destruction: National Sacrifice Areas and Native Americans." *American Sociological Review* 69, no. 4 (August 2004): 593.

Horn, Paul W. *Survey of the City Schools of El Paso, Texas*. El Paso: Department of Printing of the City Schools, 1922.

Horne, Gerald. *Black and Brown: African Americans and the Mexican Revolution, 1910–1920*. New York: NYU Press, 2005.

Horsman, Reginald. *Race and Manifest Destiny: The Origins of American Racial Anglo-Saxonism*. Cambridge: Harvard University Press, 1981.

Hovious, Jo Ann Platt. "Social Change in Western Towns: El Paso, Texas, 1881–1889." MA thesis, University of Texas, El Paso, 1972.

Huddleston, John David. "Good Roads for Texas: A History of the Texas Highway Department, 1917–1947." PhD diss., Texas A&M University, 1981.

Hugill, Peter J. "Good Roads and the Automobile in the United States, 1880–1929." *Geographical Review* 72, no. 3 (July 1982): 327–49.

Institute of Public Affairs. *Urban Renewal for Texas.* Austin: University of Texas, 1957.

Ireland, Robert E. "Prison Reform, Road Building, and Southern Progressivism: Joseph Hyde Pratt and the Campaign for 'Good Roads and Good Men.'" *North Carolina Historical Review* 68, no. 2 (April 1991): 125–57.

Jackson, Kenneth T. *Crabgrass Frontier: The Suburbanization of the United States.* New York: Oxford University Press, 1985.

Jackson, Kenneth T. "A Nation of Cities: The Federal Government and the Shape of the American Metropolis." In *Annals of the American Academy of Political and Social Science* 626 (November 2009): 11–20.

Jacoby, Karl. "Between North and South: The Alternative Borderlands of William H. Ellis and the African American Colony of 1895." In *Continental Crossroads: Remapping US Mexico Borderlands History*, edited by Samuel Truett and Elliot Young. Durham: Duke University Press, 2004.

Jacobs, Jane. *The Death and Life of Great American Cities.* New York: Vintage, 1992.

Jacobs, Jane, and the Cosmopolitan Metropolis. "2012 UHA Presidential Address." *Journal of Urban History* 39 (September 2013): 881–89.

Jones-Correa, Michael. "The Origins and Diffusion of Racial Restrictive Covenants." *Political Science Quarterly* 115, no. 4 (Winter 2000–2001): 541–68.

Juárez, Miguel. *Colors on Desert Walls: The Murals of El Paso.* El Paso: Texas Western Press, 1997.

Juárez, Miguel. "From Buffalo Soldiers to Redlined Communities: African American Community Building in El Paso's Lincoln Park Neighborhood." In "New Directions in Black Western Studies," special issue, *American Studies Journal* with *American Studies International* 58, no. 3 (2019): 107–24.

Karlsrud, Bonnie Resler. *Dale Resler Behind the Scenes in El Paso: A Historical Narrative.* Pub. by the author, 2007.

Kaszyynski, William. *The American Highway: The History and Culture of Roads in the United States.* Jefferson: McFarland, 2012.

Kelley, Ben. *The Pavers and the Paved: The Real Cost of America's Highway Program*. New York: Donald W. Brown, 1971.

Kohout, Martin Donell. "Hugh Stephenson: Pioneer Settler and Trader in El Paso." *Handbook of Texas Online*, updated September 2, 2022. http://www.tshaonline.org/handbook/online/articles/fstcx.

Kun, Josh, and Flamma Montezemolo. *Tijuana Dreaming: Life and Art at the Global Border.* Durham: Duke University Press, 2012.

Kunstler, James Howard. *The Geography of Nowhere: The Rise and Decline of America's Man-Made Landscape*. New York: Touchstone, 1995.

Kurashige, Scott. "The Many Facets of Brown: Integration in a Multiracial Society." *Journal of American History* 91, no. 1 (June 2004): 56–68.

Lang, Robert E., and Rebecca R. Sohmer. "Legacy of the Housing Act of 1949: The Past, Present, and Future of Federal Housing and Urban Policy." *Housing Policy Debate* 11, no. 2 (2000): 291–98. http://dx.doi.org/10.1080/10511482.2000.9521369.

Larice, Michael, and Elizabeth Macdonald. *The Urban Design Reader*. New York: Routledge, 2007.

Lawrence, Glenn. *Condemnation: Your Rights When Government Acquires Your Property*. New York: Oceana Publications, 1967.

Lay, Shawn. "Imperial Outpost on the Border: El Paso's Frontier Klan No. 100." In *The Invisible Empire in the West: Toward a New Historical Appraisal of the Ku Klux Klan of the 1920s*, edited by Shawn Lay. Urbana: University of Illinois Press, 1992.

Lay, Shawn. *War, Revolution, and the Ku Klux Klan*. El Paso: Texas Western Press, 1985.

Leavitt, Helen. *Superhighway-Superhoax*. New York: Doubleday and Company, 1970.

Lee, Barbara. *Renegade for Peace and Justice: Congresswoman Barbara Lee Speaks for Me*. Lanham, MD: Rowman & Littlefield, 2008.

Lewis, Robert. "Stories of Chicago: Narrative, Theory and Evidence." *Journal of Urban History* 38, no. 6 (October 2012): 1138–44.

Lewis, Tom. *Divided Highways: Building the Interstate Highways, Transforming American Life*. Ithaca, NY: Cornell University Press, 2013.

Liebschutz, Sarah F. "Neighborhood Revitalization in the United States: The Decentralization Dynamic." *Public Administration Quarterly* 14, no. 1 (Spring 1990): 86–107.

Lim, Julian. *Porous Borders: Multiracial Migrations and the Law in the US-Mexico Borderlands*. Chapel Hill: University of North Carolina Press, 2017.

Linden, Eugene. "The Exploding Cities of the Developing World." *Foreign Affairs* 75, no. 1 (January–February 1996): 52–65.

Ling, Peter. *America and the Automobile: Technology, Reform and Social Change, 1893–1923*. Manchester, UK: Manchester University Press, 1992.

Logan, Michael F. *Fighting Sprawl and City Hall*. Tucson: University of Arizona Press, 1995.

Lopez, Ronald William. "The Battle for Chavez Ravine: Public Policy and Chicano Community Resistance in Post-War Los Angeles, 1945–1962." PhD diss., University of California–Berkeley, 1999.

Mahoney, Timothy R. "The Small City in American History." *Indiana Magazine of History* 99, no. 4 (December 2003): 311–30.

Marquez, Benjamin. *Power and Politics in a Chicano Barrio: A Study of Mobilization Efforts and Community Power in El Paso*. Lanham, MD: University Press of America, 1985.

Martínez, Oscar J. *Border Boom Town: Ciudad Juárez Since 1848*. Austin: University of Texas Press, 1975.

Martínez, Oscar J. *The Chicanos of El Paso: A Case of Changing Colonization*. Southwestern Studies 59. El Paso: Texas Western Press, 1980.

Martínez, Oscar J. *Stunted Dreams: How the United States Shaped Mexico's Destiny*. El Paso: El Paso Social Justice Education Project, 2017.

Martínez, Oscar J. *Troublesome Border*. Rev. ed. Tucson: University of Arizona Press, 2006.

Massey, Douglas S., and Brendan P. Mullan. "Process of Hispanic and Black Spatial Assimilation." *American Journal of Sociology* 89, no. 4 (January 1984): 836–73.

McKay, Robert R. " Understanding Mexican Repatriation from Texas: Historical Insights." *Handbook of Texas Online*, updated April 8, 2020. http://www.tshaonline.org/handbook/online/articles/pqmyk.

McKay, Seth Shepard. *Texas Politics, 1906–1944*. Lubbock: Texas Tech University Press, 1952.

McNichol, Dan. *The Roads That Built America: The Incredible Story of the US Interstate System*. New York: Sterling Publishing, 2006.

Merken, Berry. *Wall Art*. Philadelphia: Running Press, 1987.

Metz, Leon. *Turning Points in El Paso, Texas*. El Paso: Mangan Books, 1985.

Miller, Char, and Heywood T. Sanders. *Urban Texas: Politics and Development*. College Station: Texas A&M University, 1990.

Miller, David R., ed. *Urban Transportation Policy: New Perspectives*. Lexington, MA: Lexington Books, 1972.

Miller, Loren. "The Protest Against Housing Segregation." In "The Negro Protest," special issue, *Annals of the American Academy of Political and Social Science* 357 (Jan. 1965): 73–79.

Mills, Anson, and C. H. Claudy. *My Story*. Washington, DC, 1918.

Mohl, Raymond A. *The Interstates and the Cities: Highways, Housing, and the Freeway Revolt*. Research Report. Poverty and Race Research and Action Council, 2002. http://www.prrac.org/pdf/mohl.pdf.

Mohl, Raymond A. "The Interstates and the Cities: The US Department of Transportation and the Freeway Revolt, 1966–1973." *Journal of Policy History* 20, no. 2 (April 2008): 193–226.

Mohl, Raymond A. "Planned Destruction: The Interstates and Central City Housing." In Mohl and Biles, *Making of Urban America*.

Mohl, Raymond A. "Revisiting the Urban Interstates: Politics, Policy and Culture Since World War II." *Journal of Urban History* 40, no. 5 (September 2014): 827–30.

Mohl, Raymond A. "Stop the Road: Freeway Revolts in American Cities." *Journal of Urban History* 30, no. 5 (July 2002): 674–706.

Mohl, Raymond A., and Roger Biles, eds. *The Making of Urban America*. 3rd ed. Lanham, MD: Rowman & Littlefield, 2012.

Montejano, David. *Anglos and Mexicans in the Making of Texas, 1836–1986*. Austin: University of Texas Press, 1987.

Montejano, David. "The Demise of 'Jim Crow' for Texas Mexicans, 1940–1970." *Aztlán: A Journal of Chicano Studies* 16, nos. 1–2 (1985): 27–69.

Montoya, María E. "From Homogeneity to Complexity: Understanding the Urban West." *Western Historical Quarterly* 42, no. 3 (Autumn 2011): 344–48.

Morse, Richard M. "Trends and Issues in Latin American Urban Research, 1965–1970." *Latin American Research Review* 6, no. 2 (Summer 1971): 19–25.

Mowbray, A. Q. *Road to Ruin*. Philadelphia: J. B. Lippincott, 1969.

Mullins, Paul R., and Lewis C. Jones. "Archaeologies of Race and Urban Poverty: The Politics of Slumming, Engagement, and the Color Line." *Historical Archaeology* 45, no. 1 (2011): 33–50.

Watson, Mark, ed. *The Next Exit: The Most Complete Interstate Highway Guide Ever Printed*. 11th ed. Next Exit, 2002.

Noriega, Chon, Eric Avila, Karen Mary Dávalos, Chela Sandoval, Rafael Pérez-Torres, and Charlene Villaseñor Black. *The Chicano Studies Reader: An Anthology of Aztlán, 1970–2000*. Los Angeles: UCLA Chicano Studies Research Center, 2011.

Nourse, Hugh O. "The Economics of Urban Renewal." *Land Economics* 42, no. 1 (February 1966): 65–74.

Offen, Karl. "Historical Geography II: Digital Imaginations (Progress Report)." *Progress in Human Geography* 37, no. 4 (2012): 564–77. DOI: 10.1177/0309132512462807.

Olson, R. L. *Confidential Report of a Survey in El Paso, Texas for the Mortgage Rehabilitation Division, Home Owners' Loan Corporation*. Washington, DC: Home Owners' Loan Corporation, 1936.

O'Malley, Catherine Burnside. "A History of El Paso Since 1860." MA thesis, University of Southern California, 1939.

Orndorff, Helen McCarthy. "History of the Development of Agriculture in the El Paso Valley." MA thesis, University of Texas at El Paso, 1957.

Orsi, Richard J. *Sunset Limited: The Southern Pacific Railroad and the Development of the American West, 1850–1930*. Berkeley: University of California Press, 2005.

Otero, Lydia. *La Calle: Spatial Conflicts and Urban Renewal in a Southwest City*. Tucson: University of Arizona Press, 2010.

Pandey, Gyanendra. "The Subaltern as Subaltern Citizen." *Economic and Political Weekly* 41, no. 46 (November 2006): 4735–41.

Paxson, Frederic L. "The Highway Movement, 1916–1935." *American Historical Review* 51, no. 2 (January 1946): 236–53.

Perales, Monica. *Smeltertown: Making and Remembering a Southwest Border Community*. Chapel Hill: University of North Carolina Press, 2010.

Presner, Todd, David Shepard, and Yoh Kawano. *HyperCities: Thick Mapping in the Digital Humanities*. Cambridge: Harvard University Press, 2014.

Quattlebaum, Charles B. "Military Highways." *Military Affairs* 8 (January 1944): 225–38.

Rae, John B. *The Road and the Car in American Life*. Cambridge: MIT Press, 1971.

Ramírez, Manuel B. "El Pasoans: Life and Society in Mexican El Paso, 1920–1945." PhD diss., University of Mississippi, 2000.

Ramírez, Manuel B. "When the Mexicans Go Home: El Paso and the Mexican Exodus, 1929–1934." MA thesis, University of Texas at El Paso, 1994.

Reade de Pellicano, Lilia Patricia. "*Abuelitas de El Paso*: A Selective Cultural Study." MA thesis, University of Texas at El Paso, 1989.

Reséndez, Andrés. *Changing National Identities at the Frontier: Texas and New Mexico, 1800–1850.* Cambridge and New York: Cambridge University Press, 2004.

Riddick, Winston Wade. "The Politics of National Highway Policy, 1953–1966." PhD diss., Columbia University, 1973.

Rizvi, Fazal. "Speaking Truth to Power: Edward Said and the Work of the Intellectual." In Satterthwaite, Watts, and Piper, *Talking Truth, Confronting Power.*

Rodríguez, Marisol, and Hector Rivero. "ProNaF, Ciudad Juárez: Planning and Urban Transformation." *ITU Journal of the Faculty of Architecture* 8, no. 1 (2011): 196–207.

Román, Belinda. "Cíudad Juárez-El Paso: The Formation of a Cross-Border Market: Mexico/US Economic Relations in Perspective, 1840s–1920s." PhD diss., London School of Economics and Political Science, 2003.

Romo, David R. *Ringside Seat to the Revolution: An Underground Cultural History of El Paso and Cd. Juárez, 1893–1923.* El Paso: Cinco Punto Press, 2005.

Romo, Ricardo. "Responses to Mexican Migration, 1910–1930." *Aztlán: A Journal of Chicano Studies* 6, no. 2 (Summer 1975): 173–96.

Romo, Ricardo. "The Urbanization of Southwestern Chicanos in the Early Twentieth Century." In *Chicano: The Evolution of a People*, edited by Renato Rosaldo, Robert A. Calvert, and Gustav L. Seligmann Jr. Malabar, FL: Robert E. Krieger Publishing, 1982.

Rose, Mark H. "Express Highway Politics, 1939–1956." Thesis, Ohio State University, 1973. https://etd.ohiolink.edu/.

Rose, Mark H. "Reframing American Highway Politics, 1956–1995." *Journal of Planning History* 2, no. 3 (August 2003): 212–36.

Rose, Mark H., and Raymond A. Mohl. *Interstate: Highway Politics and Policy Since 1939.* Knoxville: University of Tennessee Press, 2012.

Rothstein, Richard. *The Color of Law: A Forgotten History of How Our Government Segregated America.* New York: Liveright, 2017.

Russell, James W. "Class and Nationality Relations in a Texas Border City: The Case of El Paso." *Aztlán: A Journal of Chicano Studies* 16, nos. 1–2 (1985): 217–39.

Saldívar, Ramón. "Lyrical Borders: Modernity, the Nation and Narratives of Chicano Subject Formation." *Narrative* 1 (January 1993): 36–44.

Sandstrum, Allan W. "Fort Bliss: The Frontier Years." MA thesis, University of Texas at El Paso, 1962.

Satterthwaite, Jerome, Michael Watts, and Heather Piper, eds. *Talking Truth, Confronting Power*. Stroke on Trent, UK: Trentham Books, 2008.

Schermbeck, Clarence E. *Urban Renewal for Texas*. Austin: Institute of Public Affairs, University of Texas, 1957.

Schubert, Frank N. *On the Trail of the Buffalo Soldiers: Biographies of African Americans in the US Army, 1866–1917*. Wilmington, DE: Scholarly Resources, 1995.

Schulze, Jeffrey M. "The Chamizal Blues: El Paso, the Wayward River, and the Peoples in Between." *Western Historical Quarterly* 43 (Autumn 2012): 301–22.

Shackelford, Renae Nadine, and James Robert Saunders. *Urban Renewal and the End of Black Culture in Charlottesville, Virginia: An Oral History of Vinegar Hill.* Jefferson: McFarland, 1998.

Sheridan, Thomas E. "Embattled Ranchers, Endangered Species, and Urban Sprawl: The Political Ecology of the New American West." *Annual Review of Anthropology* 36 (2007): 121–38.

Sisneros, Samuel E. "El Paseño, Padre Ramon Ortiz: 1814–1896." *Password* 44, no. 3 (1999): 107–21.

Sixteenth Census of the United States 1940: Population, Voting Precinct 6, Part VI. Washington, DC: Government Printing Office, 1943.

Soja, Edward W. "The Socio-Spatial Dialectic." *Annals of the Association of American Geographers* 70, no. 2 (June 1980): 207–25.

Soja, Edward W. *Thirdspace: Journeys to Los Angeles and Other Real-and-Imagined Places*. Oxford: Basil Blackwell, 1996.

Soule, Gardner. *The Long Trail: How Cowboys and Longhorns Opened the West.* New York: McGraw-Hill, 1976.

St. John, Rachel. "Divided Ranges: Trans-Border Ranches and the Creation of National Space Along the Western Mexico-US Border." In *Bridging National Borders in North America*, edited by Benjamin Johnson and Andrew R. Graybill. Durham: Duke University Press, 2010.

Stein, Clarence S. "Towards New Towns for America." *Town Planning Review* 20, no. 3 (October 1949): 203–82.

Steward, Frank M. *Highway Administration in Texas.* Austin: University of Texas Bulletin 3423, 1934.

Stoddard, Ellwyn R. *The Role of Social Factors in the Successful Adjustment of Mexican American Families to Forced Housing Relocation: A Final Report of the Chamizal Relocation Research Project, El Paso, Texas*. Department of Planning Research and Development, Community Renewal Program, City of El Paso, 1970.

Strickland, Rex W. "Six Who Came to El Paso: Pioneers of the 1840's." *Southwestern Studies* 1, no. 3 (1963): 35.

Taylor, Quintard. "Comrades of Color: Buffalo Soldiers in the West, 1866–1917." In *Western Voices: 125 Years of Colorado Writing*, edited by Steve Grinstead and Ben Fogelberg. New York: Fulcrum, 2004.

Taylor, Quintard. *In Search of the Racial Frontier: African Americans in the West, 1528–1990.* New York: W. W. Norton, 1998.

Teaford, Jon C. "Urban Renewal and Its Aftermath." *Housing Policy Debate* 11, no. 2 (2000): 443–65. http://dx.doi.org/10.1080/10511482.2000.9521373.

Thompson, Marc, and Fred M. Morales. "El Paso: A Culture History and Urban Biography for the Union Plaza Redevelopment Program." Chap. 2 in *The Union Plaza Downtown El Paso Development Archaeological Project: Overview, Inventory and Recommendations*, edited by Stephen Mbuto and John A. Peterson. ARC Archaeological Technical Report no. 17. El Paso: Sun Metro Transit Authority, 1998.

Threadgill, Marguerite E. "History and Overview of McNary, Texas." *Handbook of Texas Online*, updated July 31, 2020. https://tshaonline.org/handbook/online/articles/hlm49.

Timmons, W. H. "American El Paso: The Formative Years." *Southwestern Historical Quarterly* 87, no. 1 (1983): 1–36.

Timmons, W. H. *El Paso: A Borderlands History*. Foreword by David. J. Weber. El Paso: University of Texas at El Paso, 1990.

Timmons, W. H. "El Paso, Texas." In *The Portable Handbook of Texas*, edited by Roy R. Barkley and Mark F. Odintz. Austin: Texas State Historical Association, 2000.

Tirres, Allison Brownnell. "Lawyers and Legal Borderlands." *American Journal of Legal History* 50, no. 2 (2010): 157–99. http://dx.doi.org/10.1093/ajlh/50.2.157.

Tobin, Kathleen A. "The Reduction of Urban Vulnerability: Revisiting 1950s American Suburbanization as Civil Defense." *Cold War History* 2, no. 2 (2002): 1–32.

Topalov, Christian. "Traditional Working-Class Neighborhoods: An Inquiry into the Emergence of a Sociological Model in the 1950s and 1960s." *Osiris* 18 (2003): 212–33.

Torok, George D. *From the Pass to the Pueblo: El Camino Real de Tierra Adentro National Historic Trail*. Santa Fe, NM: Sunstone Press, 2012.

Trillo, Maria Eugenia. "The Code-Switching Patterns of Rio Linda Community of El Chamizal in El Paso, Texas: An Emic Perspective of Syntactic Constraints." PhD diss., University of New Mexico, 2002.

Triplett, Henry F., and Ferdinand A. Hauslein. *Civics: Texas and Federal.* Houston: Rein & Sons, 1912.

United States National Park System Advisory Board. *American Latinos and the Making of the United States: A Theme Study.* Washington, DC: National Parks Service Advisory Board, 2013.

Utley, Robert M. *Changing Course: The International Boundary, United States and Mexico, 1848–1963.* Tucson: Southwest Parks and Monuments Association, 1996.

Vierra, Bradley J. *Keystone in Context: A Significant Archaic Period Site in El Paso, Texas.* El Paso: Keystone Heritage Park, 2009.

Villa, Raul Homero. *Barrio-Logos: Space and Place in Urban Chicano Literature and Culture.* Austin: University of Texas Press: 2000.

Weingroff, Richard F. "Federal-Aid Highway Act of 1956: Creating the Interstate System." *Public Roads* 60, no. 1 (Summer 1996): 10–11.

Weiss, Marc A. "Real Estate History: An Overview and Research Agenda." *Business History Review* 63, no. 2 (Summer 1989): 241–82.

Whitaker, Matthew C. *Race Work: The Rise of Civil Rights in the Urban West.* Lincoln: University of Nebraska Press, 2005.

Wiel, Samuel Charles. *Water Rights in the Western United States.* New York: Arno Press, 1979.

Williams, Welborn J., Jr. "The Buffalo Soldiers' Brush with 'Jim Crow' in El Paso." MA thesis, University of Texas at El Paso, 1996.

Williamson, Thad. *Sprawl, Justice, and Citizenship: The Civic Costs of the American Way of Life.* New York: Oxford University Press, 2010.

Woods, Louis Lee, II. "The Federal Home Loan Bank Board, Redlining, and the National Proliferation of Racial Lending Discrimination, 1921–1950." *Journal of Urban History* 38, no. 6 (2012): 1036.

Workable Program for Community Improvement. Washington, DC: Housing and Home Finance Agency, July 1962.

Wynn, A. O. "A Study of the Operations of the El Paso Public Schools During the School Years 1930–31 Through 1945–46." MA thesis, University of Texas at El Paso, 1946.

Zaragosa, Alex. "Recent Chicano Historiography: An Interpretive Essay." *Aztlán: A Journal of Chicano Studies* 19, no. 1 (Spring 1988–1990): 29–34.

Index

C

D

E

G

H

O

P

R

S

T

U

V

W

Y

Z